Roofing Materials & Installation

WILLIAM P. SPENCE

Sterling Publishing Co., Inc.
New York

DISCLAIMER

Photos by William P. Spence or as credited except pages:

4 **Courtesy Dura-Lock Roofing Systems Limited**

14 and 32 **Courtesy Malco Product, Inc.**

72 **Courtesy CertainTeed Corporation**

94 **Courtesy Vermont Structural Slate Company, Inc.**

114 **Courtesy Vande Hey-Raleigh Architectural Roof Tile**

148 **Courtesy Dura-Loc Roofing Systems, Ltd.**

176 **Courtesy GAF Materials Corporation**

194 **Courtesy Johns Manville Roofing Systems**

Book Design: Judy Morgan
Editor & Layout: Rodman Pilgrim Neumann

Library of Congress Cataloging-in-Publication Data

Spence, William Perkins, 1925–
    Roofing materials and installation / William P. Spence.
        p. cm.
    Includes index.
    ISBN 0-8069-9296-4
    1. Roofing—Installation—Amateurs' manuals. I. Title.
TH2431 .S66 2004
695—dc22

2003017493

2 4 6 8 10 9 7 5 3 1

Published by Sterling Publishing Co., Inc.
387 Park Avenue South, New York, NY 10016
© 2004 by William P. Spence
Distributed in Canada by Sterling Publishing
ᶜ/ₒ Canadian Manda Group, One Atlantic Avenue, Suite 105
Toronto, Ontario, Canada M6K 3E7
Distributed in Great Britain by Chrysalis Books
64 Brewery Road, London N7 9NT, England
Distributed in Australia by Capricorn Link (Australia) Pty. Ltd.
P.O. Box 704, Windsor, NSW 2756, Australia

Manufactured in China
All rights reserved

Sterling ISBN 0-8069-9296-4

Contents

1. SELECTING THE ROOFING   4
   Aesthetic Considerations 5 • Fire Resistance 6 • Wind Resistance 7 • Warranty 7 •
   Materials 7 • Roof Slope & Pitch 8 • Roofing Materials 10

2. PRELIMINARY CONSIDERATIONS   14
   Personal Safety 15 • Commonsense Safety & OSHA Requirements 16 • Storing
   Roofing Materials 23 • Estimating Materials 23 • Ventilation 26

3. ROOF REPAIRS   32
   Roof Leaks 33 • Repairing Asphalt Shingles 33 • Replacing Wood Shingles & Shakes 36 •
   Slate Roof Repairs 38 • Repairing Flashing 39

4. PREPARING THE ROOF   40
   Plywood, OSB & Waferboard Sheathing 41 • Fiberboard Roof Sheathing 42 • Underlayment
   Used On Sloping Roofs 44 • Reroofing 45 • Doing a Tear-Off Asphalt Roof 51 •
   Removing Other Roofing Materials 53

5. WOOD SHINGLES & SHAKES   54
   Cedar Shingles 56 • Cedar Shakes 57 • Sheathing 58 • Recommended Shingle & Shake
   Weather Exposure 58 • Installing Wood Shingles 58 • Flashing 63 • Installing Wood
   Shakes 69 • Other Shake Products 71

6. ASPHALT COMPOSITION SHINGLES   72
   Typical Aphalt Shingles 73 • The Roof Deck & Underlayment 75 • Installing Asphalt
   Shingles 75 • Flashing Valleys 88 • Other Flashing Installations 91 •
   Algae Discoloration on Roofs 92

7. SLATE ROOFING   94
   Slate Sizes, Grades & Textures 95 • Preparing to Install Slate Roofing 96 • Installing
   Slate 102 • Flashing 108 • Replacing Damaged Slate 112 • Ridge Vents 112

8. CLAY & CONCRETE TILE   114
   Clay & Concrete Tile Profiles 117 • Sheathing & Underlayment 118 • Preparing
   for Tile Installation 119 • Installing the Roofing Tile 123 • Cutting Tiles 136 •
   Snow Guards 137 • Flashing 138

9. METAL ROOFING   148
   Galvanic Corrosion 150 • Coatings 150 • Thermal Considerations 151 • Types of Metal
   Roofing 152 • Installing Metal Roofing Panels 156 • Flashing 167 • Working with Metal
   Shingles 172

10. BUILT-UP, MODIFIED-BITUMEN & ROLL ROOFING   176
    Considering Built-Up Roofing 177 • Hot Bitumens 177 • Cold-Applied Bitumens 178 •
    Roofing Felts 178 • Surface Aggregate 179 • Insulation Board Products 179 • The Base
    Sheet 179 • The Built-Up Roofing Membrane 179 • Modified-Bitumen Membranes 182 •
    Flashing 183 • Roll Roofing 185 • Installing Ridges & Hips 191 • Installing a Valley 191 •
    Sealing Vent Pipes 192

11. SINGLE-PLY ROOFING   194
    Syntehtic Rubber & Plastic Single-Ply Membranes 195 • Installing Single-Ply Membranes 197 •
    Using a Thermoplastic Membrane 200 • Modified-Bitumen Membranes 201 • Flashing 201

INDEX   205

METRIC EQUIVALENTS   208

Selecting the Roofing

The choice of a roofing material directly relates to the architectural design of the house. The choice of roofing is also based on what is permitted by local building codes, the slope of the roof as well as the particular climate in which it is used and the roofing material's long-term durability. As well, materials commonly used in some regions are not as widely used in other areas. This book covers roofing materials typically used on residential housing. In addition to criteria for selection of materials discussed here, later chapters cover safety procedures for taking on a roofing project, the necessary preparation from proper sheathing, underlayment, insulation, and ventilation to roof repairs and the application of specific roofing materials. New installation and the specifics of reroofing with the same material or another material are discussed, including the decision and techniques for roofing over old roofing or removing the old roofing first.

1-2 The metal roof makes a dramatic architectural statement. The colors of the siding and trim were chosen to complement this dominant motif.

The materials covered include asphalt shingles, wood shingles and shakes, slate, clay and concrete tile, metal roofing as well as built-up bitumen, roll roofing, and various available single-ply roofing membranes.

AESTHETIC CONSIDERATIONS

For new house construction or a reroofing project, the architectural design and details of the structure should help influence the choice of the roofing material. The architectural design of a residence will require a particular type of roofing material if the house is to be an authentic representation of a period style (1-1). The color and texture directly influence the overall appearance and should be a major consideration as the residence is designed (1-2). Structural factors may need to be considered in the roof framing for heavy materials, such as slate and clay tile.

1-1 This Mediterranean-style residence has a tile roof in keeping with its character. **Courtesy Mr. and Mrs. Conley Williams**

FIRE RESISTANCE

As various materials are considered for their appearance, also give attention to the **fire resistance** of the materials available. The roof is vulnerable from burning embers blowing on it from wildfires in wooded areas. Some parts of the country are especially vulnerable in this way and the homeowner should choose the most fire-resistant material possible.

The American Society for Testing and Materials (ASTM) is concerned with developing standards, testing procedures, and specifications. The standard for fire resistance in building materials is ASTM E108, *Standard Test Methods for Fire Tests of Roof Coverings*.

The actual fire resistance of roofing materials is determined by tests made by the Underwriters' Laboratories, Inc. (UL). It is a public, nonprofit safety testing laboratory. Manufacturers submit their products to the UL for testing. The test results are used to place the roofing material into four groups.

Class A roofing materials have the **highest fire-resistance rating.**

Class B roofing materials will withstand **moderate exposure** from fire that develops from sources outside the building.

Class C materials will withstand **light exposure** to fire that develops from sources outside the building.

Nonclassified materials include untreated wood shingles and shakes.

Class A, B, and C ratings are noted on the bundles of roofing material by a label (**1-3**).

1-4 Asphalt shingles have a self-sealing thermoplastic adhesive strip that bonds to the shingle above when heated by the sun.

Roof coverings of brick, masonry, slate, clay, or concrete tile or an exposed concrete roof deck are considered Class A roof coverings. The rating for metal products can vary; the manufacturer's data should be consulted. Asphalt shingles available meet Class A through Class C ratings. Unless they have been treated with a fire-retardent chemical, wood shingles and shakes are nonclassified since they do not meet any of these standards.

It should be noted that the fire-resistance rating indicated by the roofing manufacturer will sometimes depend on the **roof sheathing.** For example, a copper roofing will have a Class A rating, if the sheathing is covered with a specified fire-resistant panel and the panels are installed exactly as directed; otherwise it will rate as a Class C installation.

Local building codes specify the minimum acceptable fire-resistance rating for residential construction in your area.

As you consider various materials, note the manufacturer's claims for meeting other requirements, such as those of ASTM E282, *Air Infiltration;* ASTM E331, *Water Penetration;* and ASTM E330, *Performance Tests.*

1-3 The fire-resistance rating of a roofing material is noted on each bundle by a label. **Reprinted with permission of the Underwriters Laboratory, Inc. © 2002**

WIND RESISTANCE

Another factor to consider is the **wind-resistance rating**. The ability of asphalt shingles to withstand high winds is determined by an independent testing laboratory, such as the Underwriters Laboratory. They test the shingle's ability to resist lifting at various wind speeds. One test standard for asphalt shingles is ASTM D3161, *Standard Test Method of Wind Resistance for Asphalt Shingles*. Under this test the shingles are subjected to a continuous velocity of 60 mph for two hours. No shingle should lift during the test. Some asphalt shingles, when properly installed, are rated to withstand winds as high as 110 mph. The wind resistance of asphalt shingles is increased by using shingles that have a self-sealing thermoplastic adhesive strip above the cutouts on the exposed side of the shingle (1-4).

WARRANTY

The manufacturer's warranty is another consideration. These typically run from 15 to 25 years on asphalt shingles. The longer the warranty, the more expensive the shingle. Since the labor costs for replacing roofing materials is high, a good-quality product with expected long-term durability is the best investment. The life of a metal roof varies depending on the type of metal and

1-6 This shingle looks like slate but is a composite made from slate, a plastic resin, and glass fibers bonded under extreme heat and pressure.
Courtesy Owens Corning

the quality of the finish. A 30-year life expectancy is common. Slate roofs, if properly installed, will last 50 years or more, while treated wood shakes and shingles will last up to 30 years.

MATERIALS

As roofing materials are considered you will find a great variety of products made from a number of materials, such as wood, steel, copper, clay, concrete, organic and fiberglass asphalt-treated materials, and slate. Products with the look of these natural materials are also available made from other materials. For example, there are metal roofing products made to look like clay tile (1-5). Some fiber-cement shingles look like wood shingles. There are fiberglass asphalt shingles, ceramic tiles, and composite materials used to make shingles that look like slate (1-6).

1-5 These metal tiles resemble clay or concrete tile but are much lighter.
Courtesy ATAS International, Inc.

The wood shakes shown in **1-7** are made by bonding pieces of cedar plank under extreme pressure combined with a resin and glass fiber. The molding forms a ribbed surface, recreating the texture of wood shakes. All of these deserve careful consideration. Factors, such as the costs of the material and installation, colors available, and years under warranty, are important. Another consideration is weight. Simulated clay tile made from metal and simulated slate made from fiberglass asphalt shingles are much lighter than the real thing and influence the roof framing requirements.

ROOF SLOPE & PITCH

The slope of the roof is also a factor when choosing a roofing material. **Roof slopes** are specified by giving the number of units of vertical rise for every 12 units of horizontal run, generally based on inches of rise for a 12-inch run (**1-8**).

Pitch refers to the slant of roof that is expressed as a ratio of the **total span** of the roof to the **total rise** of the roof. For example, a building 24 feet wide that has a roof with a rise of 6 feet has a pitch of 6⁄24 or ¼ (**1-9**), but a slope of 8:12, since the rise is 8 inches for every 12 inches. Be careful you do not use pitch when you really want the slope.

1-7 These wood shakes were made by bonding pieces of cedar with glass fibers and a resin under extreme heat and pressure. They have a Class A fire-resistance rating. **Courtesy Owens Corning**

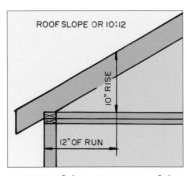

1-8 Roof slope is a ratio of the rise in units (usually inches) for a 12-unit horizontal run of the rafter.

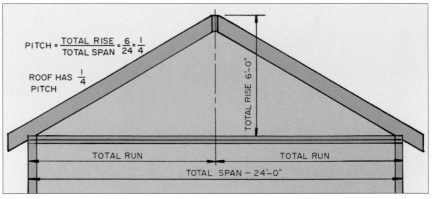

1-9 Roof pitch is a ratio of the total rise of the rafter to the total span of the roof.

LOW-SLOPE & STEEP-SLOPE ROOFS

Roofs are classified into steep-slope and low-slope types. **Steep-sloped** roofs are those that have a slope **greater** than 3 units of rise for every 12 units of horizontal run (3:12). **Low-slope** roofs are those that have a slope of 3 units of rise per 12 units of horizontal or less. The minimum acceptable slopes for various types of roofing are specified by the local building code. Following are typical examples:

Asphalt shingles are used on roofs with a slope of 2 vertical units in 12 horizontal units (2:12) or greater. For roofs with slopes between 2 in 12 and 4 in 12, a double underlayment is required.

Clay and Concrete roof tile are used on roofs with a slope of 2½ vertical units in 12 horizontal units (2½:12) or greater. For roofs with slopes between 2½ in 12 and 4 in 12 double underlayment is required.

Metal roof shingles are used on roofs with a slope of 3 vertical units in 12 horizontal units (3:12) or greater.

Metal roof panels with lapped nonsoldered seams and no lap sealant are used on roofs with a 3:12 slope. If a sealant is used the slope can be a ½:12 slope. Standing seam roofs can have a slope of ¼:12.

Mineral-surfaced roll roofing is used on roofs with a slope of one vertical unit in 12 horizontal units (1:12) or greater.

Slate and slate-type shingles are used on roofs with a slope of four vertical units in 12 horizontal units (4:12) or greater.

Wood shingles and shakes are used on roofs with a slope of three vertical units in 12 horizontal units (3:12) or greater.

Built-up, modified-bitumen, thermoplastic single-ply, and sprayed-polyurethane-foam roofing should have at least a slope of ¼ vertical unit in 12 horizontal units (¼:12).

Coal-tar built-up roofs may have a slope of ⅛ vertical unit in 12 horizontal units (⅛:12). This slope is needed to provide for drainage.

ROOF TYPES

Low-slope roofs are appropriate for some modernistic house designs but are often used where economy rather than appearance is important. Steep-slope roofs are more commonly used on quality traditional residential construction. While the slope is the designer's choice, most residences use some form of steep-slope. The three basic types of roof are the **shed, gable,** and **hip (1-10)**. The gable roof is most commonly used, with the hip roof a popular second choice. The gable roof has a central ridge with the roof sloping down from it to the wall below. The hip roof has a central ridge but it does not extend the full length of the house. It slopes to the walls on all four sides of the house. A shed roof is like a flat roof but has more slope. It extends in one direction.

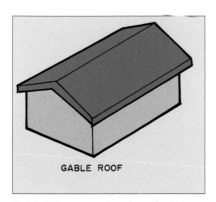
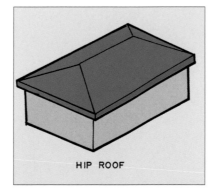
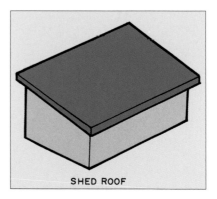

1-10 These are the three basic types of roofs used in residential construction.

Other less commonly used roofs are some variation of the shed, gable, and hip. The **butterfly roof** is actually two sloping shed roofs that meet at the center of the house. The problems with drainage and leaf collection in the valley are apparent. A more useful design is the **monitor shed roof** that sheds water rapidly and allows natural light to enter through clerestory windows (**1:11**). The **gambrel roof** has double sloping gable type roofs on each side of the house. The top roof has much less slope than the lower roof. It provides considerable space on the second floor for rooms and replaces the second floor walls of a typical two story house.

The **mansard roof** is like two sets of hip roofs with one on top of the other. As with the gambrel roof it provides considerable living space on the second floor. The gambrel roof is typical of the Dutch Colonial house, while the mansard roof is typical of the French Provincial house (**1-12**).

Some variations of the hip and gable roof are used to enclose living space when the floor plan forms a T-shape or L-shape (**1-13**).

ROOFING MATERIALS

Following are examples of some of the roofing materials available. View these materials at your local building supply dealer and pick up

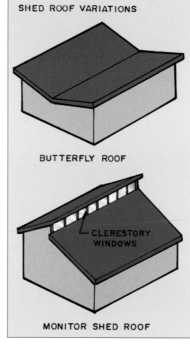

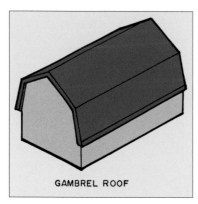

1-11 Some gable (left) and shed (right) roof variations.

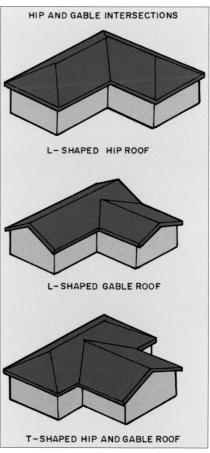

1-13 Some variations of the hip and gable roof.

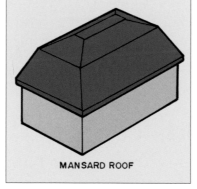

1-12 Gambrel and mansard roofs are variations of the gable and hip roof. They provide considerable living space on the second floor.

brochures giving technical and warranty information. It is helpful to know that roofing materials are estimated and sold by the **square**, which is the amount it takes to cover 100 square feet of finished roof surface, no matter what the actual surface area or shape of the particular roofing pieces. Each of these materials will be discussed in more detail in the chapters that follow.

ASPHALT SHINGLES

Two types of asphalt shingles are available. One uses a felt reinforcement made from cellulose products, such as wood and wood chips. It is referred to as an **organic type**. The felt is impregnated with asphalt. The other type uses a fiberglass reinforcement impregnated with asphalt; this is the **inorganic** type. It is the most widely used in the southern climates of North America and some are formulated for use in northern climates (**1-14**).

Ceramic-coated mineral granules are embedded in the asphalt coating saturating the base material. The asphalt coating provides a waterproof layer and the mineral granules give the shingles their color and protect them from the sun's ultraviolet (UV) rays.

WOOD SHINGLES & SHAKES

Wood shingles and shakes were the major roofing material as this country was settled. Most wood shingles available today are cut from western red cedar *(Thuja plicata)* as in **1-15**. Other woods are sometimes used but are not widely available.

Shingles are made by sawing them from the logs, so are smooth on both sides and uniform in thickness. **Shakes** are split from the log and are of varying thickness and have a rough-textured surface (**1-16**).

SLATE ROOFING

Slate roofing is cut and trimmed, generally by hand, from blocks mined at the quarry. The stone has a natural cleavage which permits it to be split cleanly along a distinct plane. It is available in a number of colors depending on the source of the stone (**1-17**). The chemical and mineral composition of a particular locality are what influence the color.

Slate roofing has a long life, is impervious to water, and is noncombustible. The mass also reduces sound penetration from exterior noise.

1-14 Asphalt shingles are available in a variety of styles and colors. **Courtesy Owens Corning**

1-16 Wood shakes have a rough, exposed surface and a varying thickness.

1-15 Wood shingles have a smooth, sawed, exposed surface.

1-17 Slate roofing tiles are fire resistant and have a long life. They provide a rough, textured appearance.

1-18 Clay roofing tiles are made in a variety of shapes and color, and are fireproof with long-term durability.

CLAY TILES

Clay tiles have been used for several thousand years and are still widely used as a roofing material in Europe. They are typically made by crushing a natural shale into a fine clay that is mixed with water and kneaded into a dough-like material. This is extruded through a die to form flat tiles; other shapes are formed by pressing the dough in a die (1-18). The tiles are then fired in a kiln.

Clay tiles are available in a range of colors developed by blending various clays. Some are colored by spraying a thin layer of a creamy clay that is baked on the tile. This layer gives the tile the desired color.

Clay tiles have a long-term durability, are impervious to water, and are noncombustible. The mass also reduces penetration of outside sounds.

CONCRETE TILES

Concrete tiles have been used in Europe for more than one hundred years but are fairly new in the United States. They are a mix of portland cement, sand, and water. The mix is extruded into molds under high pressure, thus forming the tile. It is then carefully cured.

Concrete tiles are available in a variety of colors. They are colored by adding pigment to the mix or coating the surface of the tile with a slurry of iron oxide pigment and cement (1-19). The tiles are then overlaid with a clear acrylic sealer.

The various types of concrete tile perform better in some climates than others. Before choosing, one must consult the manufacturer's data. For example, some types perform poorly in areas with extreme freezing and thawing conditions, whereas others work better in warm areas that have long periods of high humidity.

Concrete tiles have a long-term durability, are impervious to water, and are noncombustible. The mass also reduces penetration of outside sounds.

CEMENT-FIBER SHAKES & SLATE

Cement-fiber shakes and slate are made by combining portland cement and various fibrous materials. For example, one type includes organic and inorganic fibers, perlite, and iron oxide pigments producing the color. They are noncombustible and have a long life. They also resist heat and water as well as conditions of high humidity.

1-19 Concrete roofing tiles are made from portland cement, sand, and water. They are fireproof and have a long life.

METAL ROOFING

While metal roofing has been used for hundreds of years, current products are superior due to the application of modern production methods as well as technical improvements in the composition of the metal, finish coatings, and installation techniques.

Products are available simulating other materials, such as tile, wood, slate, and even asphalt shingles (**1-20** and **1-21**). They are available in the form of individual shingles and sheet products. The chapter opening photo, on page 4, shows a metal tile roofing that simulates wood shingles; it has the color of wood yet is fire-resistant and has a long life. See Chapter 9 for more information.

Metal roofing products are available in a variety of colors and are used on a wide range of architectural house styles. They have a long life and are fire-resistant. Examine the manufacturer's warranty as a choice is made. One especially

1-20 This metal roofing simulates clay tile but is much lighter. **Courtesy ATAS International, Inc.**

1-21 This metal roofing simulates laminated asphalt shingles. **Courtesy ATAS International, Inc.**

important feature to examine is the installation procedures and the method of joining the products which assure that the roof will remain watertight. When compared with other roofing materials, metal roofing provides a singular opportunity to have an installation different from those in common use. Also consider the fact that it is very lightweight, possibly reducing the cost of the roof framing required.

Metal roofing is manufactured in steel, aluminum, copper, stainless steel, and zinc.

MINERAL-SURFACED ROLL ROOFING

Mineral-surfaced roll roofing is made up of the same materials as asphalt shingles. It has an organic or inorganic base felt that is impregnated with asphalt. Then the surface to be exposed to the weather has a layer of mineral granules laid into it. This protects the felt, which prolongs the life of the product.

This is used on buildings where minimal protection is expected and long-term durability is not a major factor. The main feature is the low cost. It is installed as a prepared roof deck.

MODIFIED-BITUMEN ROOFING

Modified-bitumen roofing systems consist primarily of a polymer modified-bitumen (modified-asphalt) sheet reinforced with one or more plies of a fabric, such as polyester, glass fiber, or both. The sheets are applied to the roof deck as a one-, two-, or three-ply system consisting of a modified-bitumen membrane and a base sheet. The base sheet may be asphalt-saturated and coated organic felt or an asphalt-coated fiberglass base sheet.

This system is available in a range of modifiers and reinforcing plies designed to meet the varying requirements for residential and commercial buildings. It is used on low-slope and steep-slope roofs.

Preliminary Considerations

Working on a roof requires a knowledge of safety procedures. This is a dangerous place to work and the roofer must be constantly alert for circumstances that could lead to an accident. Since accidents do happen, the roofing contractor must have workers' compensation insurance. Accidents occur not only from falls but also during lifting operations and from tool-related injuries.

PERSONAL SAFETY

Wear proper clothing. This should be rather loose, nonbinding clothing that is durable to protect from abrasion. Work clothes get dirty from substances that will not wash off easily, so must be frequently discarded. Shoes should have soft rubber soles for good traction and less damage to the shingles.

Use shingle pads to hold the shingles so they do not slide off the roof and to protect the knees and legs, as shown opposite; it is also a good idea to wear knee pads (2-1). Finally, have a good, sturdy pair of gloves to use when lifting materials or working with hot bitumen. Even the shingles get very hot to the touch in the summer. Eye protection is needed when cutting materials or working with hot bitumen.

As you work, be constantly alert for circumstances that may cause an accident. Loose shingles or improperly placed scaffolding or ladders are common problems. Wet shingles, underlayment, and exposed

decks are a major problem; let them dry before going on the roof. Keep the roof free of loose scrap material.

Wear eye protection during tear-off jobs or when working with hot materials. If someone is working above you, wear a hard hat. Never point a power stapler or nailer at anyone. When the power stapler/nailer is not being used, shut it off so it does not go off accidently. Consider it like a loaded gun.

Store flammable materials in closed containers well away from the building. When moving shingles and other materials onto the roof, spread them out so that the load is not concentrated in one area. The deck could give way under the load.

LIFTING SAFELY

Back injuries are a common accident in roofing work. Heavy bundles of shingles have to be moved and lifted onto the roof. When lifting, observe the following recommendations:

- Keep your back straight.

- Be certain your feet are on a solid surface and spread them 15 to 18 inches apart.

- Use the muscles in your arms, shoulders, thighs, and legs to make the lift (2-2). Do not rely on the back muscles.

2-1 When working on the roof, wear knee pads (right) and use shingle pads (opposite) to keep the shingles from sliding off the roof and for leg protection.
Courtesy Malco Products, Inc.

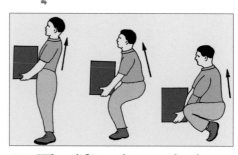

2-2 When lifting a box or other heavy bulky article, keep your back straight, be certain your feet are on a solid surface spread 15 to 18 inches apart, and lift with the muscles in the arms, shoulders, thighs, and legs.

Get someone to help lift materials you consider too heavy.

When possible, use mechanical lifts to move shingle bundles on the site and lift them onto the roof.

• If materials are carried up a ladder, break down a bundle into units that you can handle (2-3). Place the partial bundle over your shoulder, leaving one hand free to grasp the rungs of the ladder (2-4).

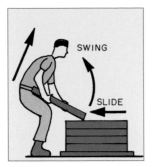

2-3 One way to lift a shingle bundle is to slide one on your upper leg and lift, keeping it pressed against your body.

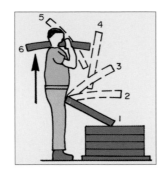

2-4 Keep your back straight; carry the bundle on your shoulder.

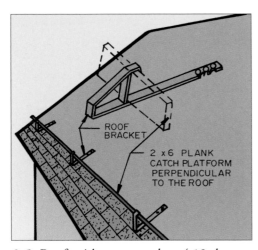

2-5 Roofs with a greater than 4:12 slope that are 16 feet or more above the ground should have a catch platform at the edge of the roof.

COMMONSENSE SAFETY & OSHA REQUIREMENTS

Fall protection regulations by the Occupational Safety and Health Administration (OSHA) for roofs undergoing new roofing installation include the requirement that, for roofs with slopes up to 8:12 and a maximum fall distance of 25 feet, 2 x 6-inch toe boards are installed on roof jacks that are secured to the roof deck at a right angle to the eave (2-5). On the steeper slopes, 6:12 to 8:12, these toe boards are to be installed at intervals not greater than 8 feet below the working area.

Roofs with slopes greater than 8:12 require the use of conventional fall protection systems such as lifelines connected to a safety harness (2-6 and 2-7), safety nets, and a guard rail system. Since these regulations change over time it is important the roofing contractor keep up to

2-6 Some roof conditions require that the roofer wear a harness and lifeline to protect against falls.
Courtesy ABC Supply Co., Inc.

2-7 A safety harness is required by OSHA regulations for work on steep roofs and under other hazardous working conditions. This unit has a retractable lifeline. **Courtesy ABC Supply Co., Inc.**

2-8 On lower sloped roofs, 2 x 4 toe boards nailed to the roof are used for foot support.

date on current requirements. State and local governmental agencies might also have safety regulations related to fall protection systems.

On lower slope roofs safety regulations may permit the use of 2 x 4s toe-nailed to the roof to provide foot support (**2-8**). They can also be installed with roof jacks (refer to **2-20**).

LADDER SAFETY

Keep current of any changes in regulations and make sure you are aware of all OSHA requirements. Following are some useful safety tips.

- Always check the ladder for defects; never use a damaged ladder.

- The feet of the ladder must be on a firm, non-slippery surface. On soft earth, place squares of ¾-inch plywood below each leg (**2-9**).

- Use fiberglass ladders; ladders made of wood deteriorate with age. An aluminum ladder is dangerous if used around electric extension cords or overhead wiring.

- Nail blocking chocks at the feet to keep them from slipping (**2-10**).

2-9 (Left) On soft soil, support the ladder with large pieces of plywood.

2-10 (Right) Use blocking secured to the ground to keep the ladder from slipping away from the house.

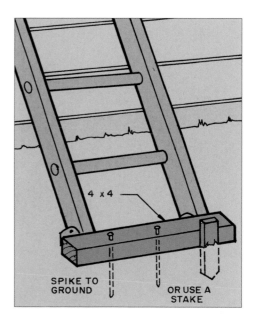

- When properly erected, the top of the ladder should extend 3 feet above the edge of the roof; the feet should be away from the wall a distance equal to ¼ the working height of the ladder (**2-11**).

- Move heavy materials to the roof with a hoist.

- Observe the maximum-load carrying capacity of the ladder. Recommended is a ladder rated 1A or 300 pounds.

- The top of the ladder is kept from sliding by securing it to eye bolts screwed into the fascia; otherwise, ladders frequently slip sideways. A reroofing job will require that the gutters are protected (**2-12**). New construction will not yet have a gutter, so the ladder can be tied to the fascia (**2-13**).

- Keep the rungs and your shoes free of mud, ice, and water.

- Avoid using ladders when exposed to high winds.

- Never use a ladder as a horizontal plank on a scaffold; it is not designed to withstand loads in this position and may fail or be damaged.

- Allow a three-foot overlap between sections of a 36-foot extension ladder and a four-foot overlap if using a 48-foot extension ladder.

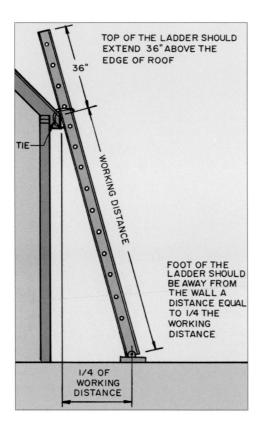

2-11 The proper way to position a straight or extension ladder.

TOP OF THE LADDER SHOULD EXTEND 36" ABOVE THE EDGE OF ROOF

36"

TIE

WORKING DISTANCE

FOOT OF THE LADDER SHOULD BE AWAY FROM THE WALL A DISTANCE EQUAL TO 1/4 THE WORKING DISTANCE

1/4 OF WORKING DISTANCE

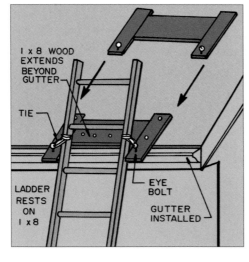

2-12 When using ladders on reroofing jobs, protect the gutter and tie the ladder to wood strips nailed on each side.

1 x 8 WOOD EXTENDS BEYOND GUTTER

TIE

LADDER RESTS ON 1 x 8

EYE BOLT

GUTTER INSTALLED

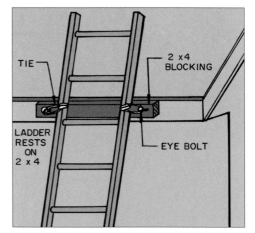

TIE

2 x 4 BLOCKING

LADDER RESTS ON 2 x 4

EYE BOLT

2-13 New construction will not have a gutter, so the ladder can rest on and be tied to blocking nailed to the fascia.

SCAFFOLDING

Scaffolding should be carefully installed and all connections fully secured. Do not use any damaged parts, such as braces, screw legs, or trusses. Metal scaffolding has the advantage of being easy to erect and move. Be certain the scaffolding will carry the loads you plan to put on it. OSHA regulations state that a scaffold should be able to carry four times the maximum load intended to be placed on it (2-14).

Scaffolds higher than 10 feet should have top and middle guard rails as well as toe boards on all open sides. The top guard rail should be 42 inches high. The distance between the scaffolding and the building must be 14 inches or less.

If the scaffold is above 10 feet or on questionable footing, tie it to the wall of the building. Check current OSHA regulations since they change occasionally.

Planks may be laminated wood or solid wood, metal, or other approved materials. Lumber is available that has been certified to meet OSHA specifications for use in scaffolding. It will have the grade stamp on it. The platform should be at least 18 inches wide. Mount the scaffold using a ladder as shown in 2-14. Do not climb up using the braces. You could fall and they could be bent and have to be replaced. Some manufactured scaffolds have ladder rungs welded into the end panels as shown in 2-15.

2-14 (Right) A typical metal scaffold. The ends are welded and connected with diagonal rods that are bolted to the ends.

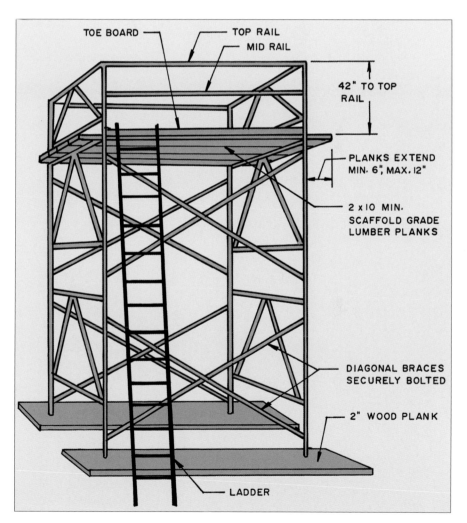

TOE BOARD — TOP RAIL
MID RAIL

42" TO TOP RAIL

PLANKS EXTEND MIN. 6", MAX. 12"

2 x 10 MIN. SCAFFOLD GRADE LUMBER PLANKS

DIAGONAL BRACES SECURELY BOLTED

2" WOOD PLANK

LADDER

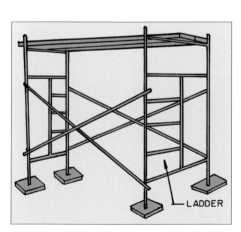

LADDER

2-15 (Above) This welded metal scaffold has ladder rungs welded into the end panels that provide access to the platform.

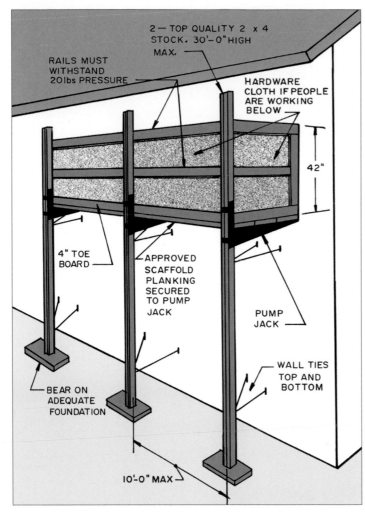

2-16 A pump-jack scaffold has a moveable platform that rides on 4 x 4-inch posts.

PUMP JACKS

Pump jacks have a moveable platform that is raised or lowered along four-inch posts (2-16). The platform is raised by stepping on the foot pedal. It can be lowered by rotating a hand-operated crank. As with scaffolding, both top and middle guard rails as well as a toe board are required. The platform must be at least 12 inches wide. Always check current OSHA requirements because these regulations do change occasionally (2-17).

A manufactured pump-jack scaffold is shown in 2-18. It uses aluminum poles and is OSHA-recognized to a height of 50 feet. The components permit rapid assembly and ease of operation.

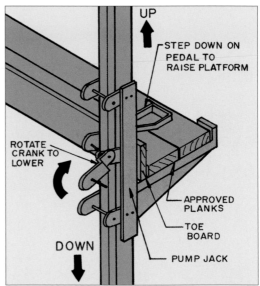

2-17 A pump jack is raised by stepping on a foot pedal and lowered by a hand-operated cranking mechanism.

2-18 (Left) This scaffolding system is provided with aluminum poles that with its components is a total system.
Courtesy Alum-A-Pole Corporation

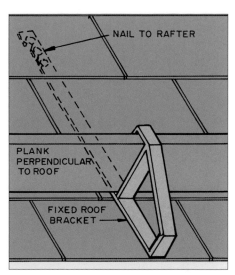

2-20 This is a fixed roof bracket that holds the plank perpendicular to the roof. It provides good foot support and keeps tools and materials from sliding down the roof.

2-19 This adjustable roof bracket is used on steep roofs. The angle of the plank can be adjusted as needed. **Courtesy ABC Supply Co., Inc.**

ROOF BRACKETS

Some roof brackets can be adjusted as needed to the slope of the roof (**2-19**); others are fixed at one orientation (**2-20**). They are nailed to the roof sheathing and the shingles are laid over them. They are generally used on roofs with a slope of 6:12 or more. They provide a fixed working surface from which the roofer can work.

Roof brackets are spaced no more than 10 feet apart. They are secured to rafters with several nails. The planks should be high-quality 2 x 10 scaffold-grade lumber. They should extend at least 6 inches beyond the bracket but no more than 12 inches. On long runs overlap the planks 12 inches on a bracket.

LADDER JACKS

Ladder jacks are metal frames that fasten to the rungs of a ladder. They extend out from the ladder to support a scaffold-grade plank (**2-21**). They are for light work and useful when working on the edge of the roof.

2-21 (Right and far right) Ladder jacks provide a quick way to support a work platform. Use approved planks and observe the manufacturer's load limits.
Courtesy ABC Supply Co., Inc. and Advanced Design Products, Inc.

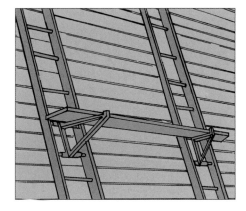

Be certain the ladders are in good condition and of high quality. The platform should be 12 inches wide. Do not space the ladders more than 8 feet apart. Tie them to the wall.

POWER LIFTS

Various types of tractors have front-end loaders or pallet forks that can be used to lift roofing materials and place them on the roof (2-22). Another type of hoist that will move roofing materials to the roof is shown in 2-23. This hoist will lift up to 400 pounds smoothly and safely to the roof. It prevents possible physical injury that might otherwise be suffered by roofers as they hand-carry shingles, builder's felt, and tools up a ladder. The hoist is portable so can be quickly moved about the construction site or to another site. Another very important application is the movement of large sheathing panels (2-24). When the platform reaches the desired height, the materials are moved along an unloading ramp to the roof deck (2-25).

2-22 Tractors with front-end lifts provide an excellent way to move materials to the roof. They are fast and safe. Observe the tractor's maximum load specifications. **Courtesy Bobcat Company**

2-23 This hoist will rapidly and safely move roofing materials to the roof. **Courtesy Safety Hoist Company**

ROOFING MATERIALS & INSTALLATION

STORING ROOFING MATERIALS

Asphalt shingles have strips of adhesive that bond the shingles together after they are installed. The heat from the sun softens the adhesive. When storing asphalt shingles put them on wood pallets in a shady spot. Cover with plastic sheeting (2-26) so they do not get wet before they are installed. Do not stack the bundles over 4 feet high. In the winter keep them in a warm place; if they get very cold, they will become stiff and hard to use and possibly crack. They should be kept above 40°F.

Store wood shingles and shakes in their original bundles in a dry place. If left outdoors, put them on wood pallets to get them off the ground and cover with plastic sheets.

Metal roofing panels are stored horizontally on wood pallets to keep them clear of the ground. Be careful they do not bend, twist, or get scratched. Cover with plastic sheeting.

Clay and concrete roofing materials are also stored on pallets and covered with plastic.

ESTIMATING MATERIALS

After a decision has been made on the type of roofing material to be used, the amounts of the various materials needed must be estimated. As noted in Chapter 1, roofing materials are estimated and sold by the **square**, which is the amount it takes to cover 100 square feet of finished roof surface, no matter what the actual surface area or shape of the particular roofing pieces. To find the number of squares, divide the roof area by 100.

2-24 (Right) Lifting large sheathing panels to the roof is a difficult job. This hoist will move them rapidly and safely.
Courtesy Safety Hoist Company

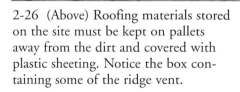

2-26 (Above) Roofing materials stored on the site must be kept on pallets away from the dirt and covered with plastic sheeting. Notice the box containing some of the ridge vent.

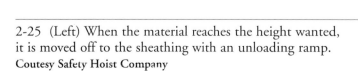

2-25 (Left) When the material reaches the height wanted, it is moved off to the sheathing with an unloading ramp.
Coutesy Safety Hoist Company

A roof contractor will most likely have a computer program into which the data for the various parts is entered and from which quantity estimates are received. Regardless of whether a computer program will be used or manual calculations made, the estimator usually visits the job site to review conditions. Small jobs can be estimated from the architectural drawings if they are available. If the house is old, on-site measurements will probably have to be taken.

One important on-site check is the condition of the rafters and sheathing. On older houses these often need replacement or repair and must be part of the overall estimate.

FIGURING THE ROOF AREA

Consider each section of the roof as a flat plane (2-27). The area of rectangular roofs is calculated by multiplying the length (A) by the width (B). Round off fractions to the next highest foot. The total square feet will be the sum of the square feet in each roof plane.

Hip roofs will have triangular areas on the hip and front roofs, as shown in **2-28**. The areas of the triangular surfaces are found by converting these into **right triangles**; a right triangle has an angle of 90 degrees (**2-29**). The area of a right triangle is found by multiplying ½ the length of the base by the height.

The hip roof area is an **isosceles triangle**; it has two sides of equal length (**2-29**). To find the area divide it into two right triangles, find the area of one of them, and multiply by 2.

To find the area of the front roof, mark off the sloped ends into right triangles, as shown in **2-28**, and calculate the area. Then calculate the rectangular area and add up the three figures.

If the roof has a dormer, measure the length and width in feet, multiply them, and subtract from the total square feet for the job. Usually chimneys, vents, and small skylights are not subtracted. The area of the dormer roof must be calculated separately and included.

Intersecting roofs will require that the area covered over one roof by the other have that area subtracted. In **2-30** a triangular area was estab-

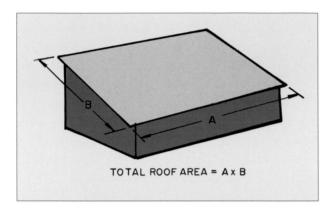

TOTAL ROOF AREA = A x B

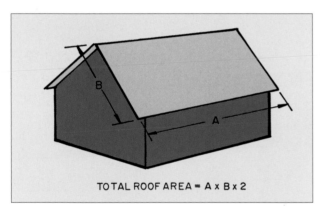

TOTAL ROOF AREA = A x B x 2

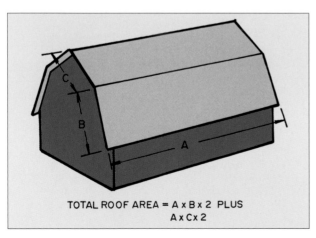

TOTAL ROOF AREA = A x B x 2 PLUS
A x C x 2

2-27 (Left top, middle, bottom) Ways to calculate roof area when all surfaces are rectangular.

ROOFING MATERIALS & INSTALLATION

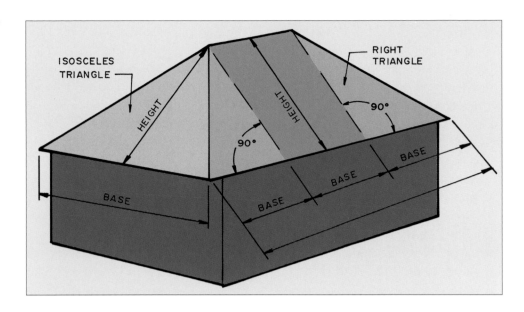

2-28 Hip roofs are made up of triangular and rectangular areas.

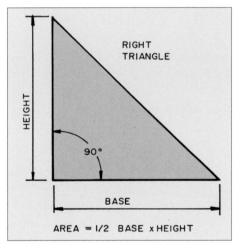
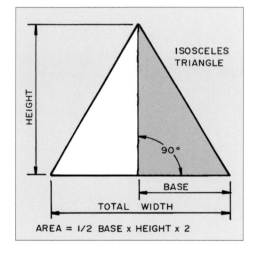

2-29 Finding the area of right and isosceles triangles.

lished. The base is the width of the intersecting building. The height is the distance from the exterior wall to the point where the ridge of the intersecting roof meets the other roof.

Once the total square footage is found, divide by 100 to find the number of squares of roofing needed, since roofing is sold by the square, covering 100 square feet, as described above. Also use this figure to order the **underlayment**.

Estimators will typically add 5 to 10 percent to the final figure to allow for waste or some unusual situation that may occur. If you run short, the shingles from the reorder may not exactly match those in the original order because they may be from a different run.

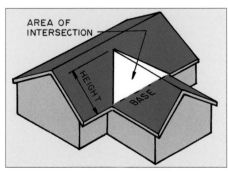

2-30 The area of intersection of two roofs can be subtracted from the total square feet of the intersected roof.

FURTHER ESTIMATES NEEDED

Now estimate the number of shingles you need for **ridge and hip caps.** An asphalt shingle can be cut to give 3 cap shingles (**2-31**), or will cover 15 lineal inches, which equals 1.25 feet. This assumes a 5-inch exposure. If the shingle to be used has a different exposure, adjust accordingly. Divide the length of the ridge or hip by 1.25 feet to find the number of asphalt shingles needed. Other roofing materials use the same method; however, take the amount of exposed shingle (such as a cedar shingle) you will use to divide the length. Some types of shingles have factory-manufactured ridge and hip shingles and give the amount to be exposed.

Another estimate needs to be made for the lineal feet of **flashing.** How this is done will depend on the flashing technique to be used. A lineal-foot estimate will give the basis needed for this calculation.

A roofing contractor will have charts that help estimate the number of **nails or staples** required. A rule of thumb used by some is to multiply the number of squares by 1.5 to get the number of **pounds** of nails needed.

VENTILATION

Roof ventilation is used to remove moisture from the air in the attic space below the roof. It will also reduce the surface temperature of the shingles. However, with today's quality products the temperature of the roofing materials is not the most damaging factor. Actually using a white shingle roof provides the lowest surface temperature. The most damaging factor is that of moisture collecting in the attic, which creates an environment for the growth of mold, and, since it wets the rafters and sheathing, will after a few years begin to rot. The moist environment also provides a home for ants and other insects.

Besides providing adequate ventilation, one very important control is in preventing moisture in the air in the living space from penetrating cracks in the ceiling and entering the attic. This control is possibly as important as ventilating the attic. Even a well-constructed house with a properly insulated ceiling can permit considerable moisture penetration. Some sources of leaks are shown in **2-32.** Seal around all light fixture boxes and pipes penetrating the ceiling and around fireplaces, access doors, and any other openings.

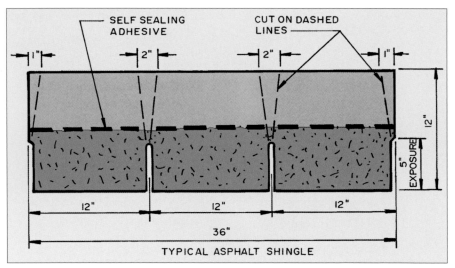

2-31 Hip and ridge caps for asphalt roofing can be cut from standard shingles. Three caps from one shingle.

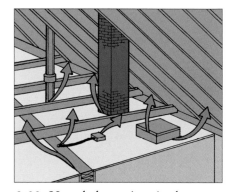

2-32 Unsealed openings in the ceiling allow air to flow into the attic. This carries the humid air into the space, thus requiring ventilation to prevent moisture from condensing on the joist, rafters, and sheathing.

Air can seep up the inside of the partitions and leak into the attic at the top plate. Covering it with a plastic sheet or caulking will help.

Now consider how to ventilate the attic; regardless of your approach you must observe the requirements of the local building code. Typically, this will require one square foot of vent area to 150 square feet of ventilated area. The ratio can be reduced to 1 in 300 if at least 50 percent, and not more than 80 percent, of the ventilating area is provided by ventilators in the upper portion of the space to be ventilated and at least 3 feet above the eave or cornice vents, with the balance of the required ventilation provided by eave or cornice vents. The net free ventilation area may be reduced to 1 in 300 if the ceiling has a vapor barrier having a transmission rate not exceeding **1 perm** installed on the warm side of the ceiling. A material having a vapor-transmission rate of 1 perm or less is considered a good vapor barrier. (The effectiveness of a vapor barrier is measured by its **perm rating**, which is the ratio of porosity of material to passage of water vapor.) The manufacturer of the venting unit used should specify the amount of venting permitted. Vents covered with wire screen have the actual vent area reduced below the size of the opening.

It is also important to have a dry crawl space or basement. Heavy moisture in these areas can infiltrate into the living area and possibly into the attic if the ceiling is not air tight.

VENTILATING A GABLE ROOF

A gable roof can be vented with soffit vents and a ridge vent, or soffit vents and gable end vents (**2-33**). These systems provide a low place (soffit vents) for air to enter and a high place (ridge vent or gable end vents) for the air to exit (**2-34**). Ridge vents are widely used (**2-35**).

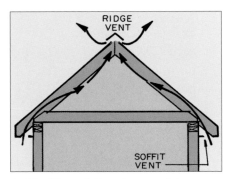

2-34 Clear airflow is essential to proper attic ventilation.

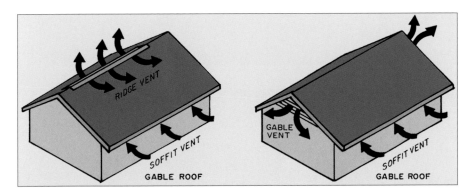

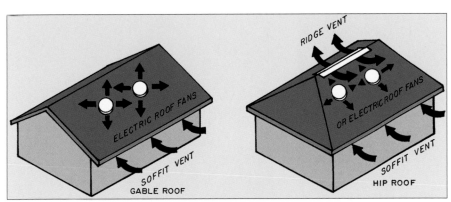

2-33 Most commonly used ways to ventilate an unheated attic.

2-35 A typical ridge vent used with asphalt shingles.
Courtesy Cor-A-Vent, Inc.

2-36 There are many architecturally pleasing ways to provide for venting the attic.

Roof venting techniques can take many forms and designs. A few examples are shown in **2-36**. These not only provide venting but serve as an architectural design feature.

The best way to get air to enter the attic from the soffit is to use continuous soffit vents (**2-37**) or perforated aluminum or vinyl soffit panels (**2-38**). Other soffit vents, such as rectangular units or ventilator plugs, often do not provide sufficient flow or even distribution of air within the attic.

It is vital when the attic is insulated that the airflow not be blocked. Usually this involves using blocking and a plywood baffle (**2-39**) or a preformed ventilation chute (**2-40**): this will hold back the insulation and prevent the airflow from being blocked by it.

Overall the design should provide an unobstructed path for air from the soffit to the ridge vent or gable end vent.

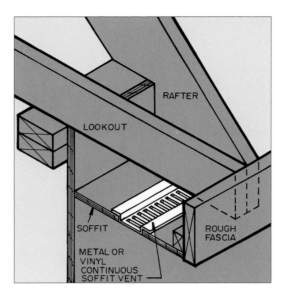

2-37 Continuous soffit vents provide excellent inflow of air to the attic.

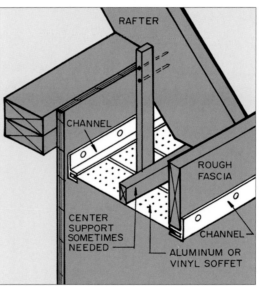

2-38 Perforated aluminum or vinyl soffit panels are one way to provide a continuous inflow of air to the attic.

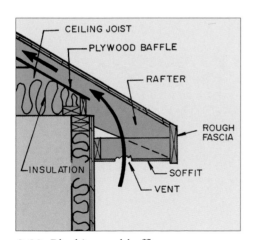

2-39 Blocking and baffles are used to compress the insulation, providing a passage for the air venting the attic.

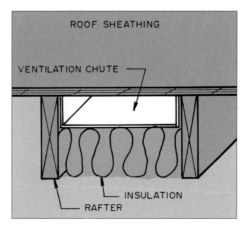

2-40 The insulation is held back with a ventilation chute, forming a clear air passageway.

CATHEDRAL CEILINGS & VENTILATION

Cathedral ceilings are popular and widely used in the living room, dining room, or den (**2-41**). They also are used when part of the attic area is converted into living space. Two situations occur in this instance. In **2-42** a partial flat ceiling is used with a steep sloped roof to form the cathedral ceiling. In **2-43** the rafters form the ceiling running all the way to the ridge. In both cases there will be insulation located between the angled room ceiling and the sheathing on the roof as well as in the vetical walls. Both instances are referred to as a cathedral ceiling.

Soffit vents provide the inflow. The space between the insulation and the roof sheathing should be at least 1½ inches to provide adequate airflow to the ridge vent (**2-44**). It is absolutely essential that this air channel be maintained and adequate air inflow at the eave be provided—otherwise moisture damage to the drywall ceiling and roof sheathing will occur.

2-41 A cathedral ceiling gives the area a lofty, open, attractive appearance.

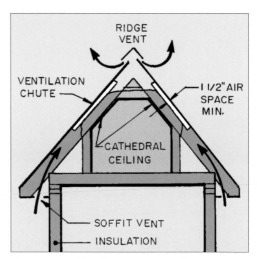

2-42 A partial flat ceiling can be made part of a cathedral ceiling when the roof has a steep slope.

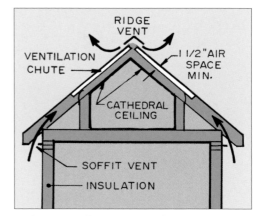

2-43 Typically, rooms in the attic space will have a cathedral ceiling run to the ridge in order to get enough headroom.

2-44 A typical cathedral ceiling on a first floor room. Notice the long air channel needed between the roof sheathing and rafter insulation.

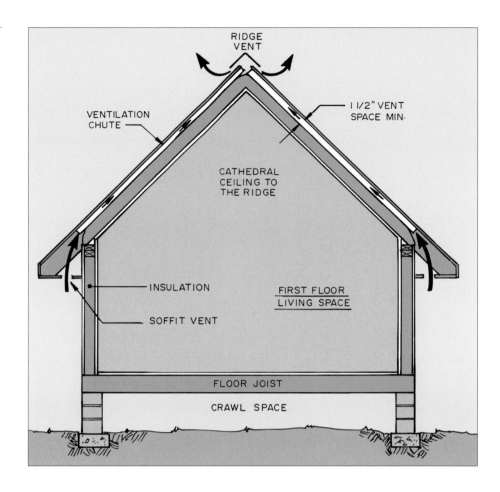

MIXING DIFFERENT TYPES OF AIR VENTS

Often a house will have ridge vents and gable louvers; however, manufacturers recommend that these not be mixed. The homeowner often likes the appearance the gable louvers give the house, but, when ridge vents are installed, the gable louvers should be closed on the inside with a piece of plywood or styrofoam sheet. The venting works more effectively if the flow from the soffit vents to the ridge vent is not interrupted by airflow from the gable vents; however, the use of gable louvers and soffit vents provides an acceptable alternate venting system.

If the roof has electric power vents and ridge vents, the flow of air is short-circuited whenever the fans turn on. When the fan is on it pulls air from the ridge vent, causing unbalanced airflow along the underside of the roof. It could also pull moisture from rain into the attic. It is best to close off air intake from power vents if ridge vents are installed.

CHAPTER 3

Roof Repairs

Over a period of years the roofing material is battered by rain, hail, snow, ice, and wind. While most materials last a long time, occasionally there is damage that should be repaired. The damage may not immediately result in a leak, so occasional visual inspection of the roof is recommended to allow for repairs that can be made before a leak occurs. Also be sure as you work on the roof to observe the safety recommendations in Chapter 2. Be careful that when you walk on the roof you do not cause additional damage. For example, some asphalt shingles become stiff and somewhat brittle as they age and freeze; it may be wise to stay off the roof in freezing weather. In any case, wear soft-soled shoes. Note that shoes can also cause damage in very hot weather.

ROOF LEAKS

If a wet spot occurs on the ceiling, go into the attic and try to find the source of the leak. Generally the leak will be so small that you will not actually be able to see light coming in from the outside.

Look for water stains on the sheathing and rafters or go into the attic when it is raining. The area of the leak typically will be moist and darker than the surrounding wood. The ceiling insulation will also be wet, giving a clue as to the approximate location. Replace the wet insulation after the leak has been repaired.

Be aware that sometimes a wet spot on the ceiling is not a roof leak but rather water from a leaking waste pipe, water pipe, or hot-water heating pipe that runs through the attic. The attic is not a good place for water pipes, especially in cold climates. Sometimes the leak may be caused by flashing that has pulled away from a wall or chimney. Check along the flashing and caulk in the area near the leak.

If it is a leak in the roofing, remember that the moisture will often run down the sheathing or rafter for some distance before it drops to the ceiling (3-1).

REPAIRING ASPHALT SHINGLES

A temporary repair can be made by sliding a waterproof sheet, such as aluminum or another shingle, between the damaged shingles (3-2). If the weather is very cold, be careful when lifting the top shingle because it may crack.

3-1 (Right) When a leak is found in the shingles, measurements can be made from the leak to the ridge or roof edge (if it is closer) and from the leak to the gable end in order to locate the hole on the outside. Some homeowners simply drive a large nail through the leak so that it protrudes above the shingles on the outside.

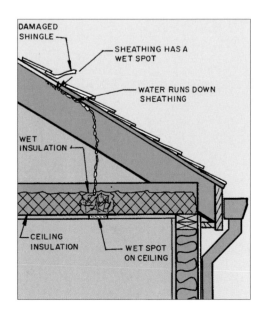

DAMAGED SHINGLE
SHEATHING HAS A WET SPOT
WATER RUNS DOWN SHEATHING
WET INSULATION
CEILING INSULATION
WET SPOT ON CEILING

3-2 An emergency repair can be made by sliding a waterproof sheet under the damaged shingle.

SLIDE A SHEET OF WATERPROOF MATERIAL UNDER THE DAMAGED SHINGLE

REPAIRING
MINOR CRACKS & TEARS

Small cracks can be repaired by lifting the shingle and laying roofing cement under it (**3-3**). Then press the shingle into the cement. Next fill the crack with cement and lay a piece of fiberglass cloth over it. This, however, does leave a rather unsightly patch that does not blend in with the color of the roofing.

REPLACING
AN ASPHALT SHINGLE

Remove the damaged shingle by first lifting the one above it and then prying out the nails with a pry bar (**3-4**). After removing the damged shingle, slide the new shingle in place (**3-5**); make certain it lines up with the adjacent shingles. If the shingles are warm, it is often possible to lift up the shingle above and nail the new one in place (**3-6**). If the shingles are cold, lift the top shingle enough to get a nail stuck into the lower one. Then drive it in place by placing a bar or other tool on the nail and hammer on the bar (**3-7**).

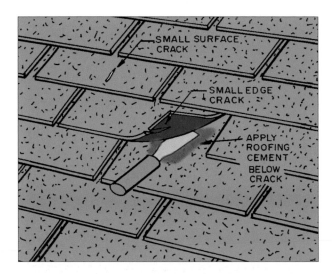

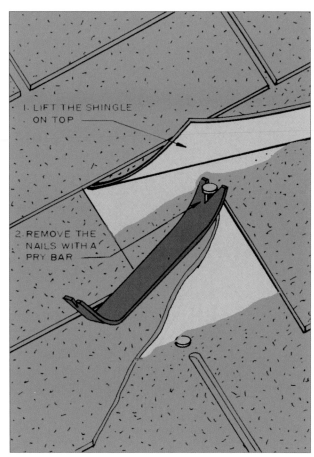

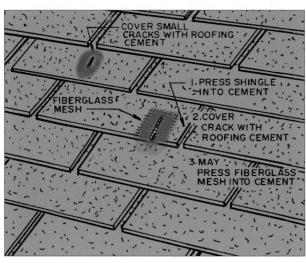

3-3 Minor cracks and small tears can be repaired by coating them with roofing cement and laying a fiberglass mesh in the cement.

3-4 Remove a damaged asphalt or fiberglass shingle by lifting the shingle above it and pulling out the nails with a pry bar. Be careful not to crack the top shingles while lifting the damaged one out.

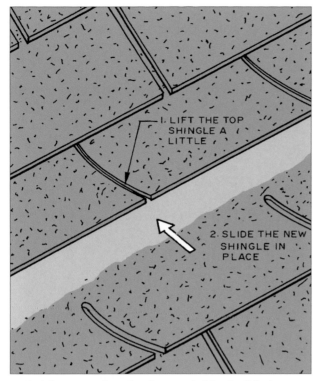

3-5 After removing the damaged shingle, lift the top shingle and slide the new one under it. Be certain the edge lines up with the other shingles in the course.

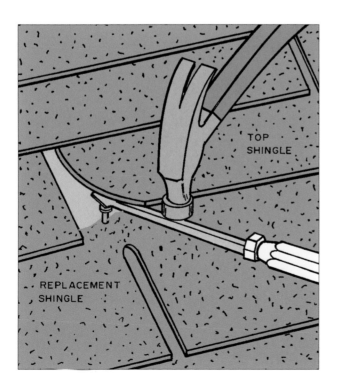

3-6 If the asphalt shingles are warm and flexible, it may be possible to lift the top shingle enough to allow a nail to be driven without cracking the lifted shingle.

3-7 (Left) If the shingles are cold and brittle, carefully lift the top shingle enough to set the nail in place with your thumb. Then set it by placing a bar or flat tool on the nail and striking it with a hammer.

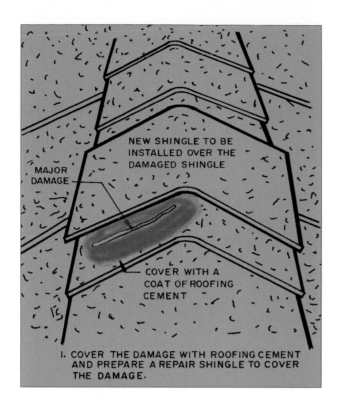

MAJOR
DAMAGE

NEW SHINGLE TO BE
INSTALLED OVER THE
DAMAGED SHINGLE

COVER WITH A
COAT OF ROOFING
CEMENT

1. COVER THE DAMAGE WITH ROOFING CEMENT
AND PREPARE A REPAIR SHINGLE TO COVER
THE DAMAGE.

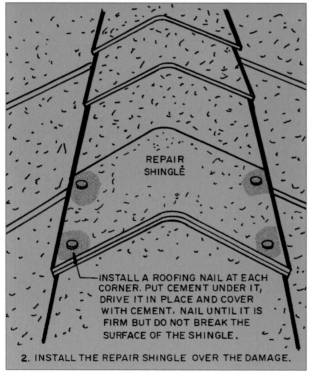

REPAIR
SHINGLE

INSTALL A ROOFING NAIL AT EACH
CORNER. PUT CEMENT UNDER IT,
DRIVE IT IN PLACE AND COVER
WITH CEMENT. NAIL UNTIL IT IS
FIRM BUT DO NOT BREAK THE
SURFACE OF THE SHINGLE.

2. INSTALL THE REPAIR SHINGLE OVER THE DAMAGE.

REPAIRING
ASPHALT RIDGE CAPS

Small cracks in a ridge cap can be repaired by covering them with roofing cement. This effectively seals them but does leave an ugly spot on the cap. Possibly it would be better to place a new cap shingle over the old one, especially if the damage is large. Apply roofing cement over the break and lay in and nail a new cap, as shown in **3-8**.

REPLACING
WOOD SHINGLES & SHAKES

Wood shingles are sawn, so are fairly smooth on both sides. Wood shakes are hand-split from sections of logs and are rough on both sides.

The steps to replace a damaged wood shingle or shake are shown in **3-9**. First drive a thin wedge under the shingle or shake on top of it. Then split the damaged shingle or shake into many small strips and pull them out. This leaves the nails still in the way, so cut them off with a hacksaw blade.

Carefully slide the new shingle in place. Tap it in until it lines up with the adjacent shingles. If it will not go all the way in, trim a little off the thin edge.

Finally, nail the new shingle by driving galvanized or aluminum shingle nails in the space between the shingles on top. The heads of the nails must rest firmly on top of the new shingle or shake, so a nail set is used to drive it home.

3-8 (Left top and bottom) Hip and ridge shingles with large cracks should be repaired by bonding and nailing a new ridge cap shingle over the damaged shingle.

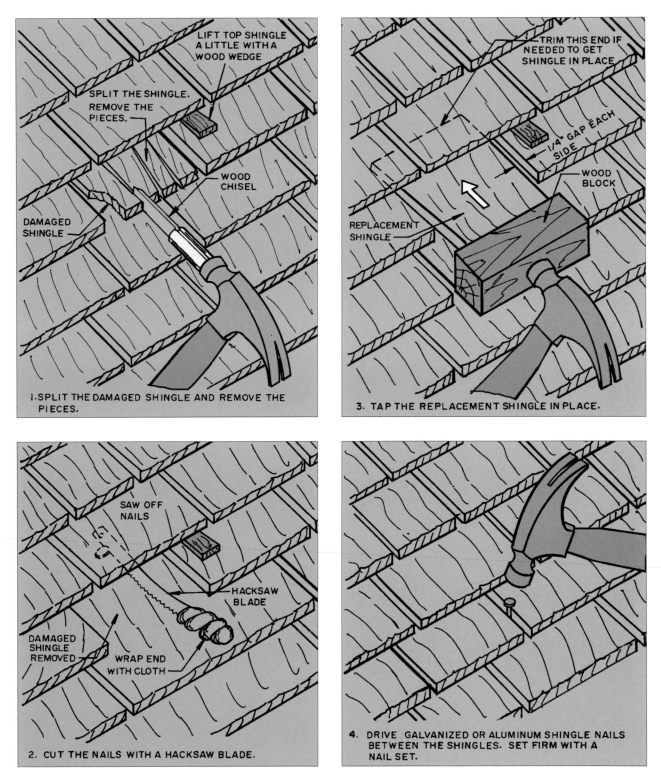

1. SPLIT THE DAMAGED SHINGLE AND REMOVE THE PIECES.

LIFT TOP SHINGLE A LITTLE WITH A WOOD WEDGE

SPLIT THE SHINGLE. REMOVE THE PIECES.

WOOD CHISEL

DAMAGED SHINGLE

3. TAP THE REPLACEMENT SHINGLE IN PLACE.

TRIM THIS END IF NEEDED TO GET SHINGLE IN PLACE

1/4" GAP EACH SIDE

WOOD BLOCK

REPLACEMENT SHINGLE

2. CUT THE NAILS WITH A HACKSAW BLADE.

SAW OFF NAILS

HACKSAW BLADE

DAMAGED SHINGLE REMOVED

WRAP END WITH CLOTH

4. DRIVE GALVANIZED OR ALUMINUM SHINGLE NAILS BETWEEN THE SHINGLES. SET FIRM WITH A NAIL SET.

3-9 Work carefully when replacing a damaged wood shingle or shake. Since they do not bend, the old shingle or shake must be split into a number of small pieces and removed. Then the nails left are sawed off and the new shingle or shake is slid into position and nailed.

SLATE ROOF REPAIRS

Begin by trying to find a slate shingle the same size and color. Color variation is common so get one fairly close to those on the roof. Some slate roofs use random-width shingles, so it is likely one will have to be cut by the dealer. You also need to identify the thickness and length. The length can usually be found by measuring a shingle along the rake. If this is not possible assume the length by doubling the exposure and adding 3 inches. Slate comes in even lengths, such as 14 or 16 inches.

Remove the remaining piece of old slate. Do this by carefully sliding a ripper (**3-10**) under the slate; hook the ripper on one of the two nails holding the slate (**3-11**). Hammer downward on the ripper. This will either cut the nail or pull it out. Do not lift up on the ripper because slates are brittle and the one above may crack. After cutting both nails, slide out the old slate.

Now drive a slate hook (**3-12**) into the exposed joint that was beneath the broken slate until it is flush with the surface (**3-13**). Slide the replacement slate up into the area covered by the old slate; then slide it down against the end of the slate hook. The hook and slate tile on each side will hold it in place (**3-14**).

Remember, slate roofs are very slippery and, if the repair is in the field above the eave, wear a fall harness. See Chapter 2 for safety information. Generally this repair is best left to the roofing contractor.

3-10 A ripper is slid under the tile and is used to cut off the nails holding it.

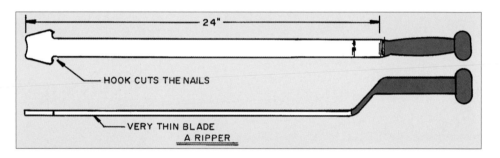

3-11 Carefully slide the ripper under the broken shingle. Hook it on a nail and drive it back, cutting the nail.

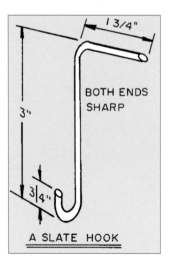

3-12 (Left) A slate hook is used to hold the new slate in place.

3-13 Drive the slate hook into the sheathing in the joint between the slates below the damaged slate.

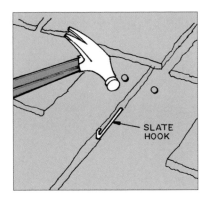

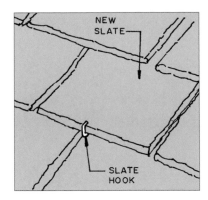

3-14 Slide the new slate into the space left after the damaged slate was removed and slide it down so that the slate hook fits over the butt.

REPAIRING FLASHING

Roof flashing is used when a pipe pierces the roof, when openings such as a skylight occur, or where a roof meets a chimney, wall, or other section of roof from a different direction. Roof flashing is discussed in detail in Chapter 5. The procedures presented show how it is installed, so look for places where it has pulled away from the area to which it was secured. It may only require some caulking to reseal the connection.

Possibly the most frequent flashing problem is with stack flashing. When this fails, replace the flashing as shown in **3-15**. Review the discussion on replacing shingles to see how to handle their removal and replacement.

3-15 (Middle, right, and far right) When stack flashing deteriorates, it is best to replace it. Temporary repairs can be made by coating it with roofing cement.

CHAPTER 4

Preparing The Roof

As the house is designed, the type of finished roofing material is selected. This choice in turn influences the design due the differences of the weight of various roofing materials and, therefore, the type and thickness of sheathing as well as the size of the rafters and other roof framing members.

Many roofing materials require a solid, fully covered surface on which the roofing material is to be applied. Solid-roof sheathing, properly installed, as shown opposite, provides the base on which the roofing is to be applied. It is essential to a satisfactory roof of long-term durability. Other roofing materials use spaced wood strips, which provide ventilation to the back of the roofing material. Building codes in earthquake-prone areas usually require solid sheathing, as do those in areas with heavy wind-driven snows or frequent rain.

The most commonly used sheathing material is in the form of 4 x 8-foot panels. Some roofers like to use 1 x 6-inch solid-wood board laid touching, forming a solid surface. Since these tend to swell and shrink as moisture in the air varies, boards wider than this will leave cracks or possibly buckle a little.

PLYWOOD, OSB & WAFERBOARD SHEATHING

Plywood panels are made by bonding wood veneers, with each layer having the grain at right angles to the adjacent layer.

OSB (oriented strandboard) panels are made by bonding strand-like wood particles arranged in perpendicular layers with a phenolic resin.

Waferboard panels are made using compressed wafer-like wood particles or flakes bonded with a phenolic resin. In some panels the wafers are arranged in layers or may be randomly or directionally oriented.

These panels should be stored on the site above the ground and covered with plastic sheets or tarps. They should be laid flat on 4 x 4-inch blocking or some other support. They are intended for structural support of the roofing material and are not expected to be exposed to the weather.

The panels are identified by the APA—The Engineered Wood Association (APA) trademark (**4-1**). It appears on panels manufactured by mills meeting the APA quality performance standards. This trademark indicates the panel is APA-rated sheathing and gives the thickness and span allowances as 32/16; this means the panel can span up to 32 inches between rafters and 16 inches between floor joists subject to predetermined loads. The exposure rating, Exposure 1, indicates it was bonded with an exterior adhesive.

Typically ½-inch thick or thinner panels are nailed with 6d ring-shank or spiral-thread nails.

For ⅝-inch panels use 8d nails (**4-2**). The nails are spaced 6 inches apart along the edges of panels which span rafters spaced less than 48 inches apart and 12 inches apart on

APA
THE ENGINEERED WOOD ASSOCIATION

RATED SHEATHING
32/16 15/32 INCH
SIZED FOR SPACING
EXPOSURE 1
000
PS 1-95 C-D PRP-108

4-1 The APA grade mark identifying rated sheathing for floors and roofs. **Courtesy APA—The Engineered Wood Association**

4-2 Roofing nails are typically obtained in large amounts, such as this 50-pound box of 6d (2-inch) roofing nails.

intermediate rafters. If the panel spans more than 48 inches, space the nails 6 inches apart on all supports (4-4). The panels are spaced ⅛ inch apart or as specified by the manufacturer to allow for expansion. They are applied perpendicular to the rafters with the end joints staggered. Aluminum H-clips are used to support the unsupported edges of the sheathing panels (4-5). If the sheathing has tongue-and-groove edges, the H-clip or other blocking is not necessary.

Sometimes spaced boards are used on roofs to be covered with wood shakes, wood shingles, concrete or clay tile, or slate to permit air circulation (4-6). Typically the roof boards are 1 x 3, 1 x 4, or 1 x 6 square-edge solid wood. Consult the manufacturers of the roofing materials for their recommendations.

FIBERBOARD ROOF SHEATHING

Fiberboard roof sheathing is a structural-load-bearing panel that provides a nailbase surface to which insulation panels, shingles, or other finish roofing materials can be applied (4-7). Panels are available prepared with a finish on the interior surface which, when used on cathedral ceilings with exposed rafters, can be painted or finished with other applied materials, such as ceiling tiles (4-8). Panel ends must bear on ¾ inches of the rafter and space ⅛ inches apart.

Fiberboard roof sheathing panels are available with a Class A fire-resistance rating. Panels are 2 x 8 feet and have tongue-and-groove edges. Nails must enter the rafter at least one inch.

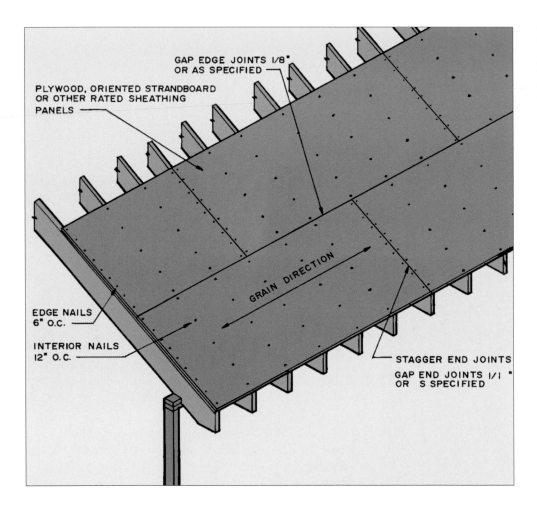

4-4 Sheathing panels are applied with the face grain perpendicular to the rafters.

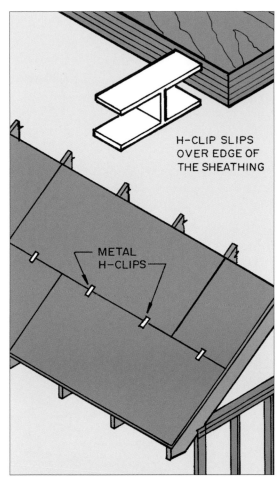

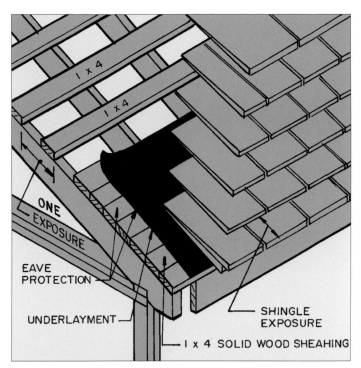

4-6 Typical installation detail for wood shingles using spaced solid-wood sheathing.

4-5 Aluminum H-clips are used to provide support on the unsupported edges of roof sheathing panels. This eliminates the need for wood blocking.

4-7 (Above) Fiberboard sheathing panels are installed in the same manner as plywood. **Courtesy Homasote Company**

4-8 (Left) Some fiberboard sheathing panels have the interior face prepared to be painted or be finished with some other material, such as ceiling tile.
Courtesy Homasote Company

UNDERLAYMENT USED ON SLOPING ROOFS

Underlayment is a waterproof material sold in rolls usually 36 or 48 inches wide. It serves as an additional layer of material protecting the sheathing from water that may penetrate the shingles or tiles. It also protects the sheathing during construction until the roofing material is applied (**4-9**).

SATURATED ORGANIC FELTS

Saturated organic felts, commonly referred to as **builder's felt**, are made using a felt composed of organic materials, such as rags, cellulose, and shredded wood. This felt sheet is then saturated with bitumen (asphalt or coal tar), and so some- times referred to as tar paper. Commonly used weights of builder's felt are 15 and 30 pounds per square, ordinarily termed 15-pound or 30-pound felt, which indicates the weight of the material per 100 square feet of roof surface (**4-10**).

SATURATED INORGANIC FELTS

Inorganic felts have a base made of fiberglass that is saturated with asphalt. There are several types available. One type is used as the base sheet when a built-up roofing system is used. Another is used under shingle and tile roofing.

Fiberglass-reinforced asphalt-shingle under- layment is available with an adhesive backing that makes it self-adhering. The top side has a granular surface coating that provides a degree of slip resistance.

4-9 The sheathing has been covered with 15-pound asphalt-saturated organic felt (builder's felt) and is ready for the application of the shingles. The roof is now weather tight.

4-10 Asphalt-saturated organic felt is available in rolls 36 inches wide. The 15-pound rolls are 144 feet long, 20-pound rolls 108 feet, and 30-pound rolls 72 feet.

SYNTHETIC ROOFING UNDERLAYMENT

Synthetic roofing underlayment has a coated woven mat utilizing advanced polymer resins. It is much stronger and lighter than 30-pound felt and will reportedly last longer after being installed.

REROOFING

As a roofing material ages and approaches the end of its useful life, a decision must be made whether to tear it off and install new materials on the sheathing or to apply the new materials over the old roofing. Reroofing decisions must be carefully made and the advantages and disadvantages considered. Some materials, such as slate, cement, and clay tiles, cannot have a new roofing material laid over them. Asphalt and wood shingles can stand reroofing over the old.

A key to successful reroofing is careful preparation of the roof deck to receive the new materials.

BUILDING CODES FOR REROOFING

Review the local building code to see what requirements must be met. The code will have a section devoted to reroofing that will have regulations related to slope and structural requirements. In addition it will limit the situations in which a new roof coating may be applied over an old roof. Following are typical of examples found in local building codes.

Reroofing over an old roof is prohibited:

- If the existing roof is water-soaked and the sheathing has deteriorated to the point it cannot serve to carry the new roof.

- If the existing roof is covered with wood shakes, slate, clay, cement, or asbestos-cement tile.

- If the existing roof has two or more coverings of any type of roofing material.

- Asphalt shingle roofs cannot be reroofed if in an area subject to severe hail damage.

REROOFING: ASPHALT SHINGLES OVER ASPHALT SHINGLES

As asphalt roofing material ages over the years, an examination will show it is cracking, curling, and in need of replacement. This brings up the decision of whether to install a new shingle over it or tear it off or start anew on the sheathing. Should you decide to reroof over the old shingles, examine the roof first to see if it has already been recovered. Building codes generally allow two layers of fiberglass or organic asphalt shingles on roofs with a 4:12 slope or less and three layers on steeper roofs. Check your local building code.

Additional roofing adds to the weight on the rafters. Typical asphalt shingles will weigh 235 pounds per square (100 square feet), so you can calculate how much additional weight is being added using your estimate of the number of squares of roofing needed.

CHECKING THE STRUCTURAL MEMBERS

Next examine the rafters and sheathing for deterioration; an examination can be made from the attic. Look for wet spots. Stick a screwdriver into suspected damage spots. If there is an area of rotted sheathing, it is necessary to remove the shingles over it and replace the sheathing. Then lay back shingles to level off the roof. If large areas are damaged, it might be best to remove all the shingles, repair the sheathing, apply new underlayment, and reshingle the entire roof.

Check the rafters in the same way. If one appears damaged, install a new rafter along the damaged one; this is called a **sister rafter**.

Check the old asphalt roofs for warped or missing shingles. Split any warped shingles and nail them down so they are flat. Replace missing shingles or you will have a sag in the new one.

The old asphalt shingles provide a cushion under the new shingles as well as an extra barrier against interior damage, especially in case high winds or heavy rains break through the new shingles. This process also eliminates the very difficult and messy job of removing and disposing of the old shingles. Finding someone to haul them away or a place to dump them can be difficult. Also, the old roof continues to protect the interior of the house should a storm come up before the new roof is in place.

INSTALLING THE NEW SHINGLES

The easiest way to reroof with asphalt shingles over old asphalt shingles is to lay new ones that match the existing pattern. Do not install new asphalt strip shingles over other types, such as hexagonal, T-lock, or the large individual shingles; a smooth finished roof will not be possible.

An underlayment is not necessary, but some roofers prefer to lay one row at the eave overlapping the drip edge.

First the ridge and hip cap shingles have to be removed. Provide a safe way to move them to the ground and store so they can be discarded. These can be easily removed with a pry bar. Slide the pry bar under and pop up the nails (**4-11**).

As you are laying the new shingles, install new flashing over any pipes (**4-12**).

4-11 (Left) Remove the ridge cap and hip cap shingles. A pry bar is a handy tool to use.

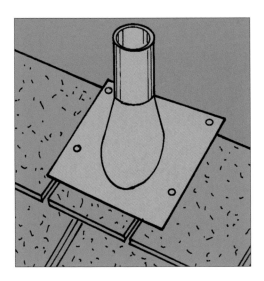

4-12 (Right) As you come to a vent pipe, install new flashing.

Some roofers install a 36-inch-wide layer of builder's felt or one of the new self-adhering underlayment sheets. This new type of sheet is adhesive-backed and fiberglass reinforced, and is used as an underlayment for asphalt shingles. It has a coarse granular surface that provides roofers with a slip-resistant surface to work on (4-13).

THE STARTER STRIP

First remove the tabs plus 2 inches or more from the top of the starter-strip shingles. What is left should equal the **exposure** of the old shingle, which is typically 5 inches. Nail the starter strips over the first course of old shingles along the eave (4-14). Place the factory-applied adhesive strip along the eaves. This butts the starter strip against the tabs of the second course. Cut 3 inches from the rake end of the starter-strip shingle so any joints will be covered by the first full course of shingles.

INSTALLING THE FIRST FULL COURSE

When installing typical three-tab shingles, many roofers prefer to begin at the rake edge with a full shingle. They then follow the pattern of the old shingles.

Cut 2 inches or more, as needed, from the butts of the shingles forming the first full course. When installed, they will fit between the butts on the third course of old shingles (4-15) and the edge of the new starter strip. Start this full-length shingle at the rake. Apply with four nails per shingle, as used on new roof construction.

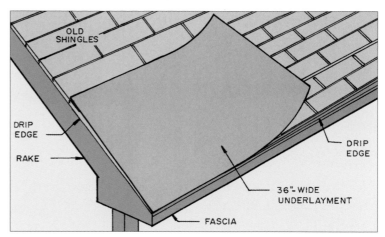

4-13 Many roofing contractors prefer to lay a 36-inch wide underlayment along the eave, especially in areas with wind, snow, and heavy rain.

4-14 Cut the starting strip to fit from the eave to the butt of the next shingle.

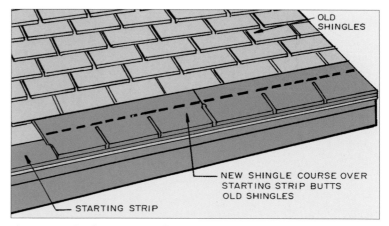

4-15 Lay the first course of new shingles over the starter strip and butt to the next course of old shingles. Cut it to width so it is even with the edge of the old shingles at the eave.

INSTALLING THE SECOND COURSE

Cut 6 inches off the first shingle and install it flush with the edge of the rake (4-16). This course will reduce the exposure of the first course to 3 inches.

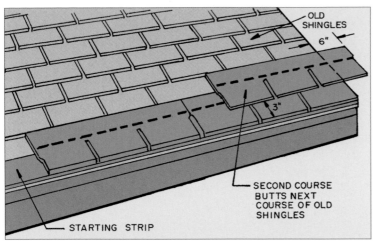

4-16 Cut 6 inches off the next shingle and install it flush on the rake. This will produce a 3-inch exposure on the first course.

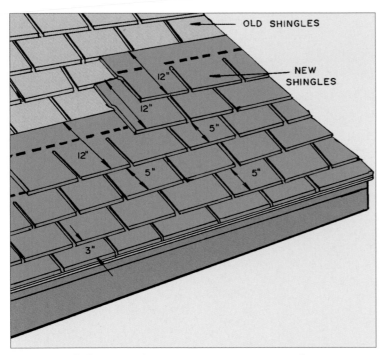

4-17 Install the succeeding courses, cutting every other one 6 inches on the rake end and butt to the old shingle above it.

INSTALLING SUCCEEDING COURSES

Install the rest of the courses by alternating the shingles at the rake. Use one full shingle on one course and one 6 inches shorter for the next course (4-17).

Note: If an underlayment is used on the eave, then the same procedure is used. The underlayment is pressed down, allowing the new shingles to go against the butt of the old shingles.

LAYING SHINGLES FROM THE CENTER OF THE ROOF

On gable roofs that are unbroken by a dormer, some roofers prefer to begin at the center of the roof and work toward each rake. Hip roofs are laid the same way (4-18). Check the layout of shingles from the centerline to see how the shingles come out at the rake. If they end with a small tab strip, move the centerline until the tabs at each rake are larger and about the same size.

After snapping a chalk line along the centerline, move over 6 inches and snap a second chalk line. This is the alignment for installing the alternate courses of shingles.

Begin by laying a 5-inch starting strip along the eave, as shown earlier in 4-14 (page 47). Proceed to lay the shingles as illustrated in 4-15, 4-16, and 4-17, except start laying from the 6-inch line set over from the centerline. Refer again to 4-18.

Nail the shingles using the standard pattern of four nails per shingle. The nails should be long enough to go through two layers of shingles and just penetrate the back of the sheathing. Typically a 1½-inch nail will be used on two layers of roofing and a 1¾-inch nail on three layers.

Finish as described for new roofs by installing the ridge and hip caps, checking and repairing the flashing, and installing ridge vents if specified.

REROOFING: ASPHALT SHINGLES OVER WOOD SHINGLES

Generally it is best to lay in beveled 1 x 4- or 1 x 6-inch strips against the butt ends of the wood shingles (**4-19**). This provides a sound nailing surface for the asphalt shingles and will give a smoother job than without the strips. A beveled board over the ridge cap provides a smooth, new nailing surface (**4-20**). If a ridge vent is to be added, the old ridge shingles are removed and the sheathing cut back to provide the opening required along the ridge. The choice of ridge vent will vary the procedure for preparing the roof to receive it. Consult the manufacturer's recommendations.

Sometimes the shingles along the rake and eave are badly damaged and should be cut back and replaced with solid-wood strips as shown in **4-21**. Then install the asphalt shingles as described for new roofs.

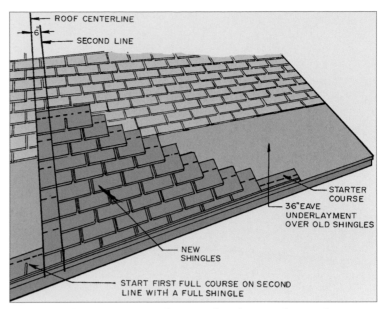

4-18 Another way to start the new shingles is to locate the center of the roof, move over 6 inches, and start there with the first row of full shingles.

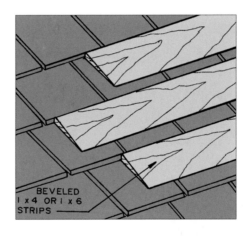

4-19 Beveled strips are used to feather out the thickness of the wood shingles and provide a nailing surface.

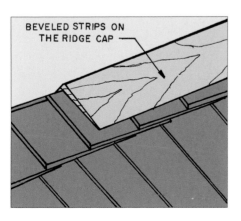

4-20 (Left) Place beveled strips to cover over the thickness of the ridge caps.

4-21 (Right) If the wood shingles on the rake and eave are damaged, cut back and replace with 1 x 6 solid-wood boards.

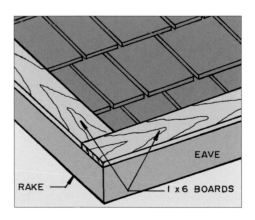

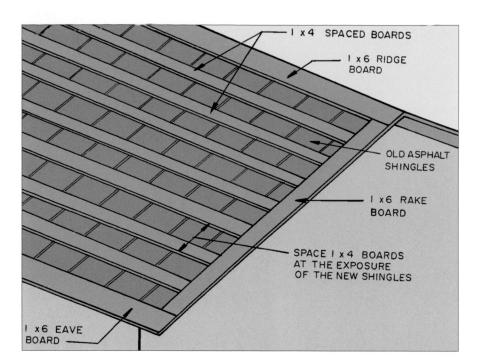

1 x 4 SPACED BOARDS

1 x 6 RIDGE BOARD

OLD ASPHALT SHINGLES

1 x 6 RAKE BOARD

SPACE 1 x 4 BOARDS AT THE EXPOSURE OF THE NEW SHINGLES

1 x 6 EAVE BOARD

4-22 Install 1 x 4-inch spaced solid-wood strips over the old asphalt shingles and 1 x 6 strips on the ridge, rake, and eave. Nail the new wood shingles to them.

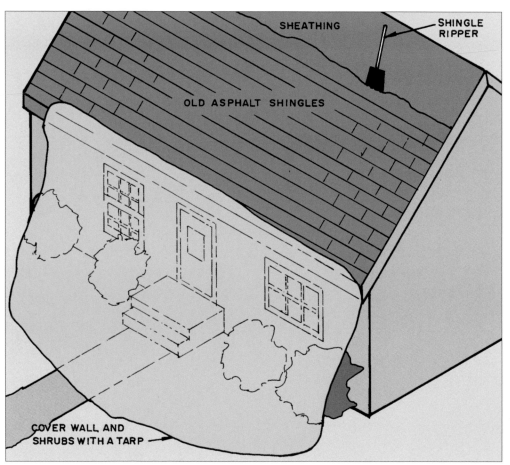

SHEATHING

SHINGLE RIPPER

OLD ASPHALT SHINGLES

COVER WALL AND SHRUBS WITH A TARP

4-23 Protect the wall, shrubs, and sidewalk before tearing off the old shingles. Have plans settled for loading and hauling away the debris.

REROOFING: WOOD SHINGLES OVER ASPHALT SHINGLES

Wood shingles can be laid over asphalt shingles but the roof must have a slope of 3:12 or greater. Check the local codes. Be certain the sheathing is sound and will hold nails. Remove the asphalt shingle ridge cap.

Nail 1 x 4- or 1 x 6-inch solid boards spaced the same distance as the exposure of the wood shingles (4-22). See Chapter 5 for more information. Install 1 x 6-inch solid wood along the rake and eave. If you have a valley install 1 x 4-inch solid-wood members along the edges of the valley. These will hold the new valley flashing.

It is also possible to install new wood shingles over old wood shingles, but carefully consider and take into account the extra weight. Examine the sheathing and rafters to see if repairs are needed. Nail down any loose shingles and reduce any warp to a level with the shingles beside it. The roof should have a slope of 4:12 or greater.

Install beveled strips along the butt ends of the old shingles as shown earlier in **4-19** (page 49). An underlayment is not required.

DOING A TEAR-OFF ASPHALT ROOF

If the asphalt shingles are in bad shape and, perhaps, the sheathing has large damaged areas, it is usually desirable but costly to tear off the old shingles, repair the sheathing, and install new shingles. A tear-off involves hard work. It is necessary to carefully plan how the shingles will be constantly removed from the roof to the ground. This causes a great deal of waste material and a mess on the ground.

Plan for a dumpster or large truck to be available to carry away the shingles. In some areas it is difficult to find a landfill or other waste disposal area that will accept this material. Before starting the job, the disposal problem must be solved.

The shingles may be moved down a chute onto the bed of a truck. However, it is not always possible to get a truck up close to the house without causing damage to the yard, sidewalk, and shrubbery. Another approach is to cover the entire side of the house and shrubs with heavy tarps (4-23).

Lay some on the ground to catch the shingles as they slide down the tarp.

Set up scaffolding, ladders, and roof brackets as needed to provide access and a safe working situation. Review safety recommendations in Chapter 2. Loose and broken pieces of shingles, along with torn builder's felt underlayment, creates a slippery surface. Wear shoes with soles that will reduce the chance of slipping.

Begin at the ridge and tear off the ridge cap (4-24), then remove any nails that tear through the shingles.

4-24 Use a pry bar to remove the ridge shingles. This pry bar was designed especially for use on shingles.
Courtesy Malco Tools

4-25 The shingle ripper pops up the shingles and is notched to go under and lift up the nails. **Courtesy ABC Supply Co., Inc.**

While some roofers use a square-edge shovel to strip the shingles, a shingle ripper is best (**4-25**). It has a flat, serrated blade that will slide under the shingle; the serrations fit under the nails. Push down on the handle to pop the nails and loosen the shingles (**4-26**).

Work carefully around flashing at walls, chimneys, and valleys. If the flashing is in good repair you will want to use it again. If it is to be replaced, pry it loose and discard.

As you work, stand on the clean sheathing. As soon as an area is cleaned, stand there and work on all sides; you are less likely to slip. Keep a broom on the roof (**4-27**) to help sweep shingles and especially the loose nails and small broken pieces to the eave.

Even the loose granules that fall off the old shingles are dangerous. They are hard and roll like marbles under your feet.

If the roof has two or three layers of shingles, remove one layer at a time. Multiple layers will produce a huge amount of debris.

Once one side of the roof has been stripped and all debris is loaded for removal, repair any damaged sheathing. Even if it is solid, go over and renail wherever necessary. At this point a section of the roof is clear but exposed to the weather. Cover it with an underlayment so that it is watertight. Keep a tarp handy in case rain starts before the underlayment is in place. Then proceed to remove the shingles on another side of the roof.

Remove the tarp from the wall and move the truck, if used, from the chute. Even when you are careful there will be some debris on the ground to be picked up. Finally roll a magnet (**4-28**) over the area to pick up nails and any other metal pieces that are loose. Do not forget to roll the driveway just in case something fell off during removal.

Now install new underlayment and shingle as explained in the chapters in this book on asphalt, wood, and other types of roof covering.

4-26 Start at the ridge and work toward the eave. Slide the shingle remover under the shingles and push down to pop them loose. **Courtesy Malco Products, Inc.**

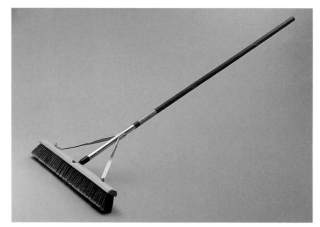

4-27 This is a heavily reinforced broom suitable for the hard use it will get on the roof. **Courtesy Malco Products, Inc.**

4-28 This rolling magnet is used to pick up all the nails that ended up on the lawn, sidewalk, and driveway. **Courtesy Malco Products, Inc.**

REMOVING OTHER ROOFING MATERIALS

If removing wood shingles, shakes, metal panels, slate, or concrete tiles, use a crowbar or pry bar to dislodge a few at a time. Work several courses, again beginning at the ridge. The problem of disposing of slate and concrete tiles is greater than asphalt shingles because of the weight. Removing wood shingles or shakes over spaced roofing presents a different set of problems. Consider placing tarps in the attic over the ceiling joists to catch debris that may fall between the sheathing strips. The attic insulation must be protected.

CHAPTER 5

Wood Shingles & Shakes

Wood shingles and shakes often have a warranty good for 20 years or more. The length of the warranty depends on whether and how they are treated. Pressure-treated shingles and shakes will have the longest warranty. Shingles and shakes that are not pressure-treated at the factory can have a wood preservative applied after they have been installed. They can also be sprayed with a fire-retardant chemical. These have to be reapplied every few years.

Wood shingles and shakes provide excellent insulation and are several times more effective than other types of roofing.

It is important to remove regularly any leaves or pine needles that are on the roof. While shingles and shakes are made from naturally rot-resistant woods, especially western red and eastern white cedar (but also pine, cypress, and redwood), the leaves or pine needles that collect on them hold moisture and reduce the life of the roofing. If the roofing is kept dry, rot is prevented and the growth of algae, moss, and fungi is reduced. The removal of algae is discussed at the end of Chapter 6. Other growth is handled in the same way. Use a chemical recommended by the local roofing contractor.

5-2 As wood shingles age, they darken and take on tones of gray and brown.

Since wood shingles and shakes are rigid, they resist uplift by the wind and depressions by hailstones. If they do dent, the depression will most likely disappear because the wood fibers will expand to their original size.

When they are newly installed, they have a light tan color (5-1). As they age they darken and take on a gray and brown tone of varied intensity (5-2). They provide a high-quality, weathered appearance.

5-1 Newly installed wood shingles, above and opposite, have a bright tan color.

WOOD SHINGLES & SHAKES

CEDAR SHINGLES

Cedar shingles are accurately sawn on both sides and present a smoother appearance than shakes (5-3). They are available in 16-, 18-, and 24-inch lengths. They are used on roofs with a 3:12 slope or greater. They are available in Grades 1, 2, and 3.

Fire-retardant cedar shingles are pressure-impregnated with fire-retardant polymers. They are available with Class C and Class B fire-resistance ratings.

CEDAR SHINGLE GRADES

As shingles are cut from wood blocks, the place at which they are cut from the log influences the grain exposed and the grade of the shingle. The types of grain are shown in 5-4.

Cedar shingles and shakes are graded following the standards of the Cedar Shake and Shingle Bureau.

Following are the grades for cedar shingles. Each bundle of shingles is identified by a blue, red, black, or green label.

Number 1 Grade, Blue Label
Clear heartwood, 100 percent edge grain and no defects; thickness of a 16-inch shingle is 5/2, meaning that the thickness of 5 shingles at the thick butt end is 2 inches. The thickness of an 18-inch is 5/2¼ and 24-inch is 4/2. For roofs 3:12 and greater. Best grade.

Number 2 Grade, Red Label
Limited sapwood and flat grain are allowed. Limited knots and defects are allowed above the clear portion. For roofs 3:12 or greater. Middle grade.

Number 3 Grade, Black Label
Unlimited sapwood and flat grain allowed. Limited knots and defects above clear portion. For roofs 3:12 or greater. Lowest grade.

Undercoursing Grade, Green Label
Unlimited defects, flat grain, and sapwood are allowed. These are used for the underlying starter course at the eaves that are not exposed to the weather.

5-3 Sawed cedar shingles are smooth on both sides and give a smooth, finished appearance.

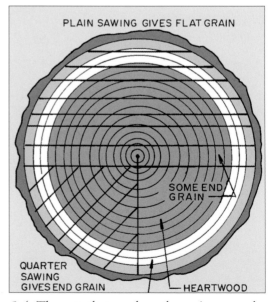

5-4 The cuts that produce the various wood shingle grains.

CEDAR SHAKES

Some types of cedar shakes have both surfaces sawn like a shingle; however, they are not as precisely manufactured as shingles. They have thicker butts than shingles and give the roof a more textured appearance. They are available in 18- and 24-inch lengths. Roof exposure is 8½ inches for the 18-inch length and 11½ inches for the 24-inch size. They are used on roofs with a 4:12 slope or steeper. They are available pressure-treated, which gives them a 30-year warranty. They are available in Grades 1, 2, and 3.

Hand-split and resawn shakes provide a rough finished surface and a heavy shadow line. The exposed face is the rough textured surface produced by splitting them from the log. The back face is smoother because it is sawn (5-5). They are available in 18- and 24-inch lengths. Roof exposures are 7½ inches for 18-inch and 10 inches for 24-inch shakes. They are used on roofs with a slope of 4:12 or greater.

Fire-retardant cedar shakes are pressure-impregnated with fire-retardant polymers. They are available with Class C and Class B fire-resistance ratings.

Following are the grades for taper-sawn cedar shakes:

Number 1 Grade, Blue Label
Face 100 percent clear; flat grain 10 percent maximum; cross grain minimum allowed. For roofs 4:12 or better. Best grade.

Premium Grade
Some Number 1 Grade except 100 percent edge grain. For roofs 4:12 or better.

Number 2 Grade, Red Label
Lower half of face is clear; flat and cross grains allowed; tight knots and other limited defects in top half of shake. For roofs 4:12 or better.

Number 3 Grade, Black Label
Face sound and serviceable; flat and cross grains allowed; limited knots and defects allowed on the entire face. Not recommended for roofs of occupied buildings.

5-5 Handsplit and resawn cedar shakes have a thick butt and provide a heavy shadow line.

TABLE 5-1 CEDAR SHINGLE & SHAKES' EXPOSURE TO THE WEATHER
(for Roofs with a Slope of 4:12 or Greater)

SHINGLES	
Length	Exposure to the Weather
16"	5"
18"	5½"
24"	7½"
SHAKES	
Length	Exposure to the Weather
18"	7½"
24"	10"

SHEATHING

Shakes and shingles are applied over spaced sheathing, which is usually 1 x 4- or 1 x 6-inch (nominal) solid softwood lumber. Review Chapter 4 for details. They may also be applied over solid sheathing, such as plywood or oriented strandboard (OSB). Solid sheathing is usually required by building codes in areas where high winds, snow, and rain are common or in seismic regions. In this condition some roofers prefer to install an underlayment and nail the 1 x 4 spacers to the sheathing. This provides some ventilation for the back of the shingles or shakes.

Solid sheathing is also used under treated shakes and shingles. In areas with high humidity and high rainfall, such as along the east coast and gulf states of the U.S., use spaced sheathing and shingles or shakes that have been pressure-treated with a wood preservative.

Fire-retardant shingles and shakes must be installed over ½-inch-minimum plywood or other approved sheathing to get a Class B fire-resistance rating. Check the local codes. They may require the sheathing be a fire-retardant-treated material.

RECOMMENDED SHINGLE & SHAKE WEATHER EXPOSURE

In **Table 5-1** are the recommended exposure to the weather for cedar shingles and shakes used on roofs with a slope of 4:12 or greater. For lower slopes consult the manufacturer. It takes special techniques for slopes less than 4:12.

INSTALLING WOOD SHINGLES

Shingles may be installed over 1 x 4- or 1 x 6-inch (nominal) solid-wood spaced boards. If 1 x 4 boards are used, they are spaced the same distance as the exposure of the shingles (**5-6**). If 1 x 6 boards are used on a shingle roof, the boards are spaced two exposures, as shown in **5-7**.

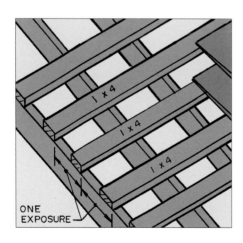

5-6 (Left) Note that 1 x 4-inch spaced wood sheathing for wood shingle installation is spaced one weather exposure.

5-7 (Right) Note that 1 x 6-inch spaced wood sheathing for wood shingle installation is spaced two weather exposures.

The sheathing is laid solid at the eave so it extends 12 to 24 inches inside the wall line. In cold climates with heavy snows this could be extended to 36 inches (**5-8**). This is covered with 30-pound builder's felt or a self-adhering synthetic polymer roofing underlayment.

Solid sheathing is also used under wood shingles. It is usually plywood or OSB (oriented strandboard). These provide a strong, smooth surface. Solid sheathing is required by codes in areas where there is a certain level of seismic risk or if fire-retardant treated shingles are used. The sheathing is covered with builder's felt or a self-adhering synthetic polymer roofing underlayment (**5-9**). See Chapter 4 for more information.

Some roofing contractors prefer to lay spaced 1 x 4 or 1 x 6 wood boards over the solid sheathing (**5-10**). This provides a small open area so that the bottom of the shingles can have ventilation.

5-9 The sheathing is being covered with a self-adhering synthetic polymer roofing underlayment. As it is rolled out over the sheathing, the protective cover over the adhesive side is removed. **Courtesy ABC Supply Co., Inc. and Tamko Roofing Products**

5-8 (Left) The sheathing is laid solid to the eave and covered with an underlayment. This provides eave protection from heavy snows and rain.

5-10 (Right) Some contractors prefer to lay spaced wood strips over solid sheathing to give the bottom of the shingles some breathing room.

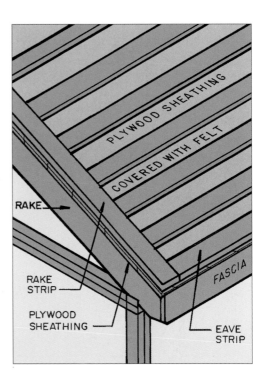

WOOD SHINGLES & SHAKES

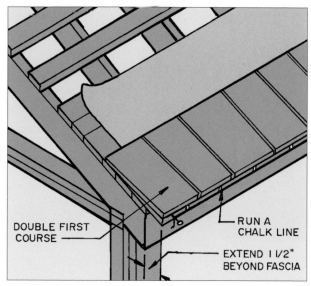

5-11 The first course along the eave must be doubled or tripled. Run a chalk line to get the butts perfectly straight.

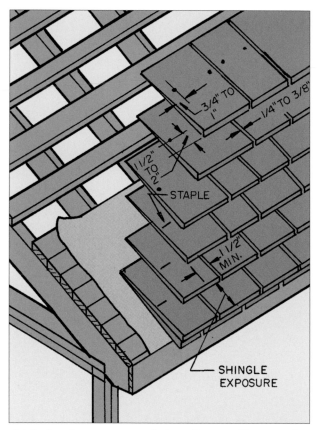

5-12 Spacing recommendations for installing wood shingles.

APPLYING WOOD SHINGLES

The first course at the eave must be doubled or tripled. The butts should extend 1½ inches beyond the fascia (5-11) and ½ to ¾ inch from the rake board. The first course can be lined up by running a chalk line from the first shingle to one temporarily nailed on the other side.

Lay the courses so that joints in any one course are separated at least 1½ inches from joints in the adjacent courses. In any three courses no two joints should be in direct alignment. Keep the spaces between adjacent shingles ¼ to ⅜ inches (5-12).

Flat-grain shingles that are wider than 8 inches should be split into two pieces and each nailed separately.

Each shingle is secured with **two nails**. They are located ¾ to 1 inch from the edge of the shingle and 1½ to 2 inches above the butt of the next course. Drive the nail firmly against the surface but do not break the wood fibers.

Staples are driven in the same location as nails and will be parallel with the butt. They are driven flush with the surface but should not break into the wood.

5-13 This roofing coil nailer with a cartridge will drive galvanized coil roofing nails ranging in length from ⅞ to 1¾ inch. It will fasten an entire bundle of shingles without reloading. **Courtesy Stanley Bostitch Corporation**

TABLE 5-2 NAIL SIZE RECOMMENDATIONS

SHINGLES — NEW ROOFS		
Length	Type	Size
16" and 18"	Box	3d (1¼")
24"	Box	4d (1½")

SHAKES — NEW ROOFS		
Length	Type	Size
18" Straight split	Box	5d (1¾")
18" and 24" Handsplit and resawn	Box	6d (2")
24" Tapersplit	Box	5d (1¾")
18" and 24" Tapersawn	Box	6d (2")

Courtesy Cedar Shake and Shingle Bureau

NAILS

Each shake or shingle is fastened with **two corrosion-resistant** nails. These include aluminum, stainless steel, or hot-dipped zinc-coated steel. If the shakes or shingles are treated, consult the manufacturer for nail recommendations. The minimum recommended nail sizes for lengths of shingles and shakes are in **Table 5-2**.

In some cases wood shingles and shakes can be nailed with a power coil nailer (5-13). Check the maximum size nail it will drive to see if it will be suitable for the application. In some cases, such as with thick wood shakes and multiple layers of shingles or shakes, nails 2 inches or longer are needed. These will have to be driven by hand.

STAPLES

When staples are used to fasten shakes or shingles, they must be **aluminum or stainless-steel** (type 304 or 316), 16 gauge with ⁷⁄₁₆-inch minimum crowns. They are driven parallel with the shake butt. They should be long enough to penetrate the sheathing by ½ inch. There will be cases where power driven staples will not be long enough so that hand driven nails will be required.

WEATHER EXPOSURE

Recommended weather exposure is shown in **Table 5-1**. As the courses are laid, it is important to be certain they are accurately laid out for each course. This can be done using a shingle hatchet. The hatchet has a series of holes into which a pin is inserted to set the weather exposure (5-14). After laying four or five courses, run a chalk line to check the alignment. Some prefer to run a chalk line to line up each course as it is laid.

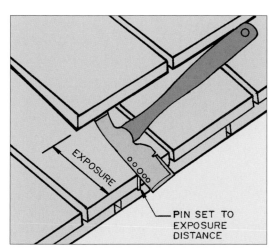

5-14 The weather exposure can be laid out with a shingle hatchet. **Courtesy ABC Supply Co., Inc.**

5-15 This hip is covered with wood shingles forming a weathertight joint.

5-16 The shingles have been laid to the hip and are being cut off in preparation for installing the hip cap.

5-17 This ridge is ready to have the shingles cut back so that the ridge cap can be installed.

HIP & RIDGE INSTALLATION

The intersections of roofs at the ridge or a hip are covered with **caps** (**5-15**).

As the shingles are laid to the hip or ridge, they overlap it and are cut back after they have been nailed in place (**5-16** and **5-17**).

The ridge and cap shingles have the edges cut back with a bevel as shown in **5-18**. As they are installed, the cap joint is alternated from side to side on every other shingle. The weather exposure is the same as that used on the roof shingles. The caps may be made on the job or purchased ready-made from the shingle supplier. Again, remember to make them with the alternating joints. The nails used are longer than those used to install the shingles. They should be long enough to allow them to penetrate the sheathing at least ½ inch.

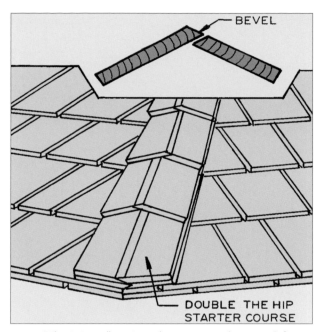

5-18 The joints forming the caps are alternated from side to side on every other shingle.

FLASHING

The most likely place that a roof is going to leak is where water flows against a vertical surface or down a valley. The most common locations are at a chimney, around a vent pipe, or where wall and roof meet.

Galvanized-steel flashing should be prepainted on both sides with a bituminous paint or other paint made especially for metal. Aluminum and copper are also good flashing materials. The choice may depend on local climatic conditions. The experience of others with these various metal flashings over a long time period provides a good guide for making a choice.

ROOF VALLEY CONSTRUCTION & FLASHING

When two roofs meet, a valley is formed and considerable water is channeled down it during inclement weather (**5-19**). The metal valley flashing has a center crimp that helps control the flow of water entering from each side.

5-19 This valley flashing has been painted on both sides and has a crimp down the center to help control the flow of water.

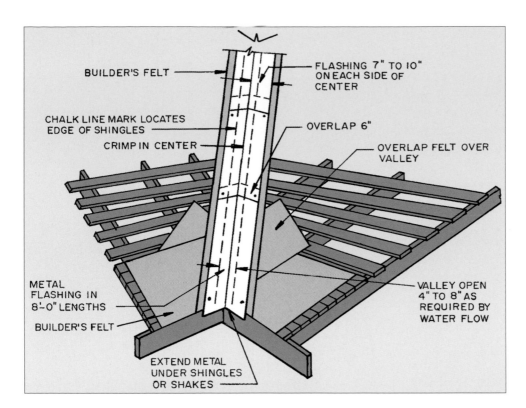

BUILDER'S FELT

FLASHING 7" TO 10" ON EACH SIDE OF CENTER

CHALK LINE MARK LOCATES EDGE OF SHINGLES

CRIMP IN CENTER

OVERLAP 6"

OVERLAP FELT OVER VALLEY

METAL FLASHING IN 8'-0" LENGTHS

BUILDER'S FELT

VALLEY OPEN 4" TO 8" AS REQUIRED BY WATER FLOW

EXTEND METAL UNDER SHINGLES OR SHAKES

5-20 Line the valley with an underlayment and install the metal flashing over it.

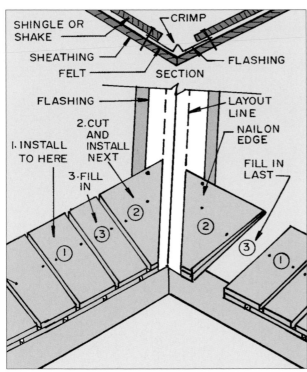

SHINGLE OR SHAKE

CRIMP

SHEATHING

FLASHING

FELT

SECTION

FLASHING

LAYOUT LINE

NAIL ON EDGE

2. CUT AND INSTALL NEXT

1. INSTALL TO HERE

FILL IN LAST

3. FILL IN

5-21 The shingle that fits over the flashing is cut to fit along the layout line and has an edge joint that does not open onto the flashing.

When installing wood shingles on roofs with slopes of 12:12 or greater, make the valley flashing at least 14 inches wide. For slopes less than this, use 20-inch-wide flashing. Begin the installation by laying 15-pound or heavier builder's felt or a synthetic underlayment that uses advanced polymer resins and woven construction (5-20). Mark lines on the flashing that locate the edges of the shingles. The width of the exposed valley is usually 4 inches, though it may be wider in areas with regular heavy rainy seasons. Some roofers recommend running the layout line so that the exposed valley flashing is somewhat wider at the eave than the top of the valley. This tapering helps carry the increasing water flow that develops toward the eave as the valley widens.

As the shingles approach the valley, install them in the order shown in 5-21. The shingle that overlaps the flashing must be cut to fit along the layout line. The edge between that shingle and the next shingle should never open onto the flashing. The overlap of the flashing must have a

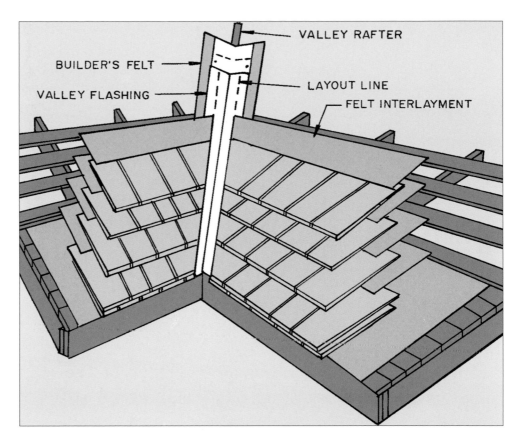

5-22 A typical valley flashing detail for wood shingles and shakes.

VALLEY RAFTER

BUILDER'S FELT

VALLEY FLASHING

LAYOUT LINE

FELT INTERLAYMENT

solid shingle above and below. The grain of the shingles must never be parallel with the center-line of the valley flashing.

Some roofers prefer to cut and fit the shingle on the flashing layout line and working away from this shingle, install the others.

A complete valley flashing detail is shown in **5-22**. Before each course of shingles is laid, an underlayment is laid to the valley and trimmed on the layout line. It can be bonded to the flashing with roofing cement. When installing the last shingle over the edge of the flashing, keep all nails near the edge of the flashing.

VALLEY/RIDGE
SADDLE FLASHING

When two intersecting roofs form a valley, and one roof is lower than the other, a lead saddle is laid over the connection and down over the flashing on the valleys on each side (**5-23**).

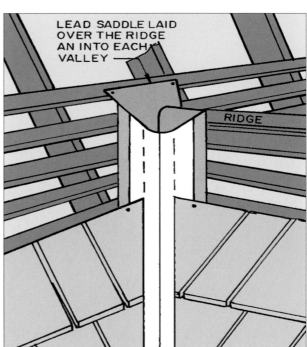

LEAD SADDLE LAID OVER THE RIDGE AN INTO EACH VALLEY

RIDGE

5-23 Lead saddle flashing covers the junction of a ridge butting a higher roof.

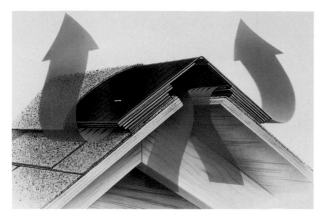

5-24 This ridge vent provides a free flow of air and can be covered with wood shingles.
Courtesy Cor-A-Vent, Inc.

5-25 This soffit vent provides the flow of air needed to make the ridge vents effective.
Courtesy Cor-A-Vent, Inc.

5-26 This vent unit is used to ventilate a roof that butts against a vertical wall. **Courtesy Cor-A-Vent, Inc.**

RIDGE VENT & FLASHING

Wood shingle and shake roofs can be vented as described in Chapter 2. The use of a ridge vent provides an efficient and attractive way to provide ventilation (**5-24**). A soffit ventilation system is shown in **5-25**. These are also widely used on hip roofs. One especially interesting venting unit is shown in **5-26**. It is used to vent a roof that butts a vertical wall.

FLASHING A CHIMNEY

As with most projects, there are several ways a job may be accomplished. The following is one commonly used method for flashing a chimney.

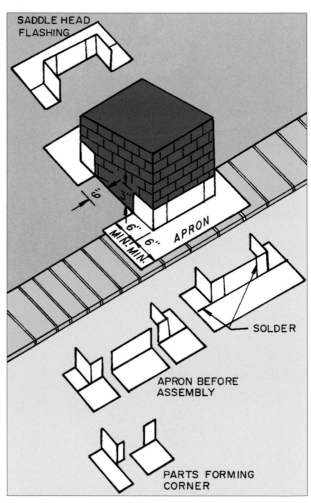

5-27 The upslope and downslope sides of the chimney are flashed with preformed aprons.

5-28 (Left and far left) Some use a cricket on the upslope side instead of a saddle apron.

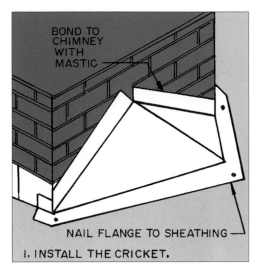

BOND TO CHIMNEY WITH MASTIC

NAIL FLANGE TO SHEATHING

1. INSTALL THE CRICKET.

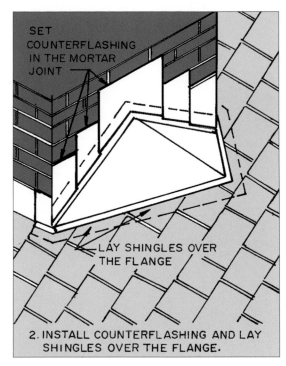

SET COUNTERFLASHING IN THE MORTAR JOINT

LAY SHINGLES OVER THE FLANGE

2. INSTALL COUNTERFLASHING AND LAY SHINGLES OVER THE FLANGE.

A preformed, soldered apron flashing is placed on the downside and a saddle flashing is put on the top or upslope side (5-27). Some roofers prefer to use a **cricket** on the upslope side (5-28). It should extend under the shingles or shakes 8 to 10 inches and up on the chimney 3 inches. It can then be covered with counterflashing.

The sides of the chimney are covered with step flashing (5-29). It is placed so the lower edge is flush with the butt of the shingle to be placed on top of it. As each shingle is placed, another step flashing unit is placed on it.

After the step flashing is in place, remove the mortar between the bricks above the step flashing and install the counterflashing (5-30).

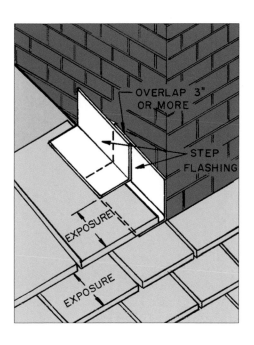

OVERLAP 3" OR MORE

STEP FLASHING

EXPOSURE

EXPOSURE

5-29 (Left) Step flashing is placed on each course of shingles along the chimney. It is set flush with the butt of the shingle.

5-30 (Right) Counterflashing is set into a mortar joint and overlaps the step flashing.

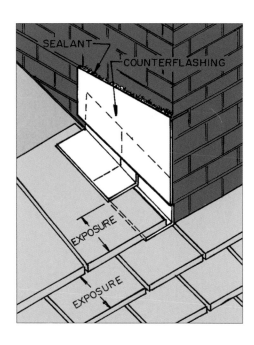

SEALANT

COUNTERFLASHING

EXPOSURE

EXPOSURE

Overlap sections of counterflashing at least three inches. A descriptive illustration is shown in **5-31**. Insert the counterflashing 1½ inches into the mortar joint. The recommended sizes of step and counterflashing are shown in **5-32**. A typical finished installation is shown in **5-33**.

FLASHING A VENT PIPE

Shingle up to the pipe and place the vent flashing over it and the shingle (**5-34**). Then fit the next course around it, cutting as necessary. Leave a space around the pipe and the downside open to allow debris to wash away (**5-34**). Some roofers install a layer of felt above the pipe and a wide shingle at the pipe to eliminate any problem a joint at this location may cause, as shown in **5-34**.

FLASHING AT A WALL

Details for flashing a roof when it meets a front or side wall are shown in **5-35**. When the sloped roof meets a wall, step flashing is used as described for chimney flashing. Instead of using counterflashing, the finished siding is laid over the step flashing. It is cut so that it is one inch above the surface of the roof. When the roof meets a wall that is parallel with the run of the shingles, an apron is used. The metal corner flashing is soldered to a long apron strip. Building felt and siding are laid over the apron, again ending one inch above the shingles.

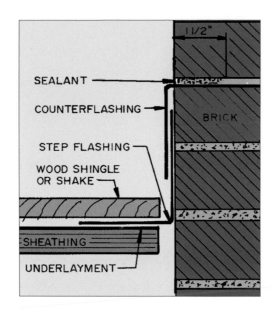

5-31 A detail showing the relationship between the step and counterflashing.

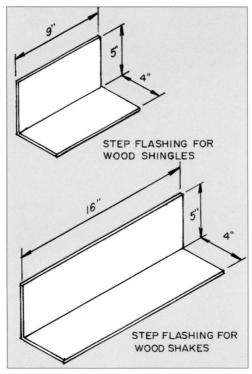

5-32 Recommended sizes for step and counterflashing.

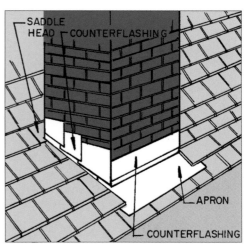

5-33 A typical finished flashing installation for a chimney.

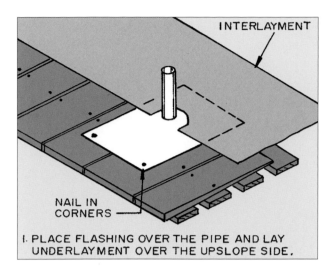

INTERLAYMENT

NAIL IN
CORNERS

1. PLACE FLASHING OVER THE PIPE AND LAY
UNDERLAYMENT OVER THE UPSLOPE SIDE.

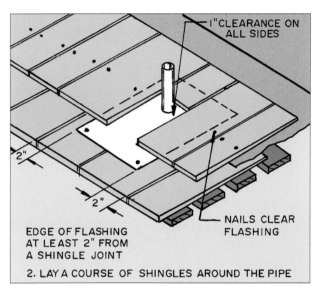

1" CLEARANCE ON
ALL SIDES

2"

2"

EDGE OF FLASHING
AT LEAST 2" FROM
A SHINGLE JOINT

NAILS CLEAR
FLASHING

2. LAY A COURSE OF SHINGLES AROUND THE PIPE

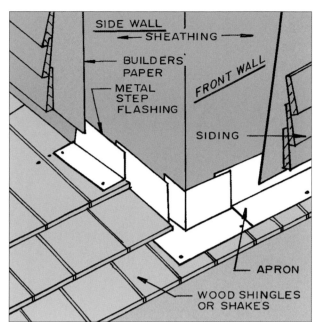

SIDE WALL
SHEATHING

BUILDERS
PAPER

METAL
STEP
FLASHING

FRONT WALL

SIDING

APRON

WOOD SHINGLES
OR SHAKES

5-35 Step flashing is used to flash the junction between a roof and a side or front wall.

INSTALLING WOOD SHAKES

Wood shakes are installed in much the same way as wood shingles. However, they are installed over 1 x 6 solid-wood spaced sheathing. The boards are spaced equal to one weather exposure (5-36) but never more than 7½ inches for 18-inch shakes or 10 inches for 24-inch shakes.

5-34 (Top left, middle and bottom left) Shingle up to the vent pipe and place the vent flashing on top of the shingle. Install the interlayment and lay the shingles around the vent pipe, leaving a clear space around it and on the downslope side.

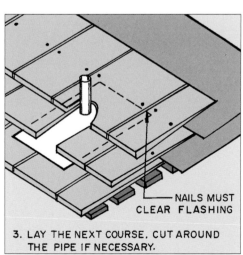

NAILS MUST
CLEAR FLASHING

3. LAY THE NEXT COURSE. CUT AROUND
THE PIPE IF NECESSARY.

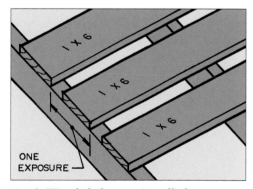

1 X 6

1 X 6

1 X 6

ONE
EXPOSURE

5-36 Wood shakes are installed over 1 x 6-inch solid-wood sheathing. They are spaced one exposure from front edge to front edge.

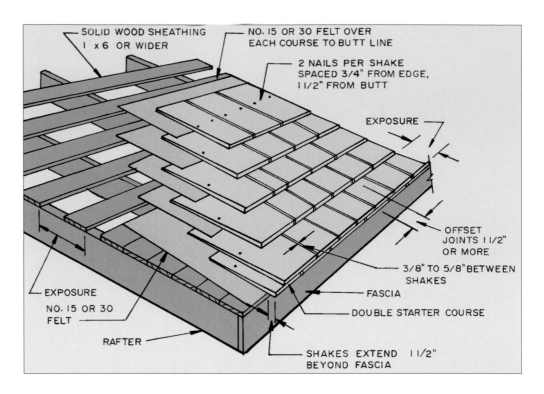

5-37 A typical installation detail for a wood shake roof.

SOLID WOOD SHEATHING
1 x 6 OR WIDER

NO. 15 OR 30 FELT OVER
EACH COURSE TO BUTT LINE

2 NAILS PER SHAKE
SPACED 3/4" FROM EDGE,
1 1/2" FROM BUTT

EXPOSURE

OFFSET
JOINTS 1 1/2"
OR MORE

3/8" TO 5/8" BETWEEN
SHAKES

FASCIA

DOUBLE STARTER COURSE

EXPOSURE
NO. 15 OR 30
FELT

RAFTER

SHAKES EXTEND 1 1/2"
BEYOND FASCIA

Solid sheathing may be used in areas where high winds, snow, and rain are common. A typical installation detail is shown in **5-37**. The starter course may be one or two courses of

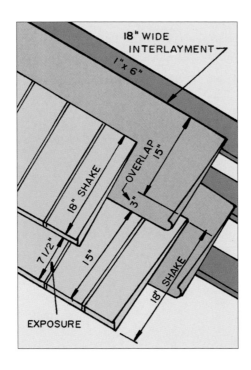

5-38 Interlayment installation with an 18-inch wood shake.

18" WIDE
INTERLAYMENT

1" x 6"

18" SHAKE

OVERLAP
15"

3"

7 1/2"

15"

18" SHAKE

EXPOSURE

cedar shingles or shakes overlaid with a course of the exposed shake. The butts extend 1½ inches beyond the fascia. An 18-inch-wide strip of 15- or 30-pound felt interlayment is laid over the top area of each course of shakes. Consult the local code for the weight of felt required. The bottom edge of the felt falls above the butt of the shake a distance equal to twice the weather exposure. For example, an 18-inch shake installed with a 7½-inch exposure would have the felt applied 15 inches above the butt (**5-38**). This felt will cover the top 3 inches of the shake and extend 15 inches onto the sheathing. The top edge of the felt must rest on the spaced sheathing.

The space between shakes should be ⅜ to ⅝ of an inch. The joints between shakes should be offset at least 1½ inches from the joint on the course below.

Hip caps and ridge caps are installed in the same manner as for wood shingles.

5-40 This wood shake is made from recycled wood products, reinforced vinyl, and cellulose fiber.
Courtesy Re-New Wood

...n attic
... Courtesy

...er a ridge
...st be long
...st ½ inch.
...cribed for

CTS

...ducts that
...are made
...stic. One
...ade from
...inyl, and
cellulose fiber. It has a Class A fire-resistance rat-ing and a Class D impact rating, which are the highest achieved from Underwriters Laboratories. It has a 50-year warranty and is available in pieces 22 inches long. Information about other shingle and shake products also appear in other chapters in this book covering other materials.

Asphalt Composition Shingles

Asphalt composition shingles are the most widely used roofing material. They are easy to install, are available in a range of colors, textures, shapes, and types (**6-1**), and have a long life. The mixing of two shapes of shingles on a roof can be seen in the photo opposite. Asphalt composition shingles are available with a Class A fire-resistance rating and can be installed to resist lifting by high winds. Since the roof is a major part of the overall design of the house, the selection of the type of asphalt shingle and the color are major considerations.

Asphalt composition shingles are available with organic or inorganic base material. See Chapter 1 for more information.

6-1 These light-tan, laminated-asphalt shingles provide a heavy shadow line and an overall pleasing appearance.

TYPICAL ASPHALT SHINGLES

Manufacturers have a range of sizes and shapes of asphalt shingles available. The most frequently used types are shown and described in **Table 6-1**, on page 74. The **laminated self-sealing shingle** is composed of two layers, with the cut out top layer exposing a layer below. This feature produces a very attractive shingle and a slight shadowline, enhancing the overall appearance (**6-2**).

One type of **multitab shingle** has a staggered butt line that produces a similar appearance but is not laminated; the **three-tab shingle** is the one most commonly used (**6-3**). The **no-cutout square-tab shingle** provides a smoother overall appearance. **Individual locking shingles** are available in several designs; they provide excellent resistance to strong winds.

6-2 The laminated-asphalt-composition shingle produces an attractive textured surface and provides an interesting shadow line.

6-3 The three-tab asphalt-composition shingle is the most widely used roofing material. **Courtesy Owens Corning**

TABLE 6-1 TYPICAL ASPHALT SHINGLES

Product	Configuration	Approx. Shipping Weight per Square (pounds)	Shingles per Square	Bundles per Square	Width (Inches)	Length (inches)	Exposure (inches)	ASTM Fire & Wind Ratings
Laminated self-sealing random tab shingle	Various edge, surface texture, and application treatments	240–360	64–90	3–5	11½–14¼	36–40	4–6⅛	Class A or C fire rating. Many wind resistant.
Multi-tab self-sealing square tab strip shingle	Various edge, surface texture, and application treatments	240–300	65–80	3–4	12–17	36–40	4–8	Class A or C fire rating. Many wind resistant.
Multi-tab self-sealing square tab strip shingle	Three-tab or four-tab	200–300	48–80	3–4	12–13¼	36–40	5–5⅝	Class A or C fire rating. Many wind resistant.
No-cutout self-sealing square tab strip shingle	Various edge and surface texture treatments	200–300	65–81	3–4	12–13¼	36–40	5–5⅝	Class A or C fire rating. Many wind resistant.
Individual interlocking shingle (basic design)	Several design variations	180–250	72–120	3–4	18–22¼	20–22½	n/a	Class A or C fire rating. Many wind resistant.

Courtesy Asphalt Roofing Manufacturers Association

6-4 These laminated-asphalt shingles are bonded together by a row of self-sealing adhesive located above the cutouts.

SELF-SEALING ASPHALT SHINGLES

Self-sealing asphalt shingles have a series of beads of thermoplastic adhesive that bond the shingle above, providing resistance to wind damage (refer to **Table 6-1**, above). They seal when the shingles are exposed to normal heating by the sun (**6-4**).

THE ROOF DECK
& UNDERLAYMENT

Success requires a strong, stable roof deck. It can be a structural-rated sheathing panel or other nonveneer panel approved by the local code. See Chapter 4 for details.

The underlayment can be 15-pound builder's felt, which is an asphalt-saturated felt product. The quality of this product varies depending on its composition (**6-5**). Underlayments with an organic felt reinforced with fiberglass are strong and reliant, indicating high quality. Refer also to Chapter 4 for additional information.

INSTALLING
ASPHALT SHINGLES

The installation begins with the **drip edge** along the eave. The drip edge protects the wood fascia from moisture. Unroll the builder's felt along the eave and place it **over** the drip edge. It must go **under** the flashing along the rake (**6-6**). Nail the drip edge to the eave sheathing. Nail the rake flashing through the underlayment.

Install the underlayment parallel with the eave. Secure it with nails having large heads or better still a plastic disk (**6-7**). Do not staple the underlayment.

6-5 This 15-pound asphalt-saturated felt underlayment is secured with plastic-headed nails. The spacing is a bit wider than some recommend but the large washer heads hold better than roofing nails.

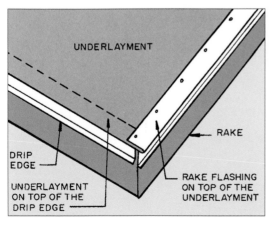

6-6 (Left) The underlayment goes on top of the drip edge on the eave and under the rake flashing.

6-7 (Right) The underlayment is secured to the sheathing with nails having plastic washers.

Space nails about 6 inches apart around the perimeter and about one inch from the edge, as shown in **6-8**. Install two rows in the interior of the sheet 12 inches from each edge. Stagger the nails in each row. Overlap pieces that meet on the end by 6 inches and double nail them by staggering the nails. Do not let two end laps fall in a line; stagger them about 12 inches apart. Edge laps from succeeding layers are overlapped 2 inches and secured with a single row of nails spaced 6 inches apart. Remember, the flashing on the rake goes **over** the top of the underlayment.

When the roof butts against a vertical wall, lap the underlayment 4 inches up on the wall (**6-9**). Cover the ridge and hips with an underlayment that extends at least 12 inches on each side (**6-10**).

In areas where icing will occur along the eaves, a self-adhering eave-and-flashing membrane is installed running at least 24 inches inside the exterior wall (**6-11**). The membrane should overlap the metal drip edge ½ to ¾ inch. It can have a 2-inch edge lap only over the part of the roof that extends outside the building. In very northern climates local codes may require the membrane to extend 36 inches inside the exterior wall.

The self-adhering membrane is an asphalt-coated, fiberglass-reinforced underlayment with one side coated with an adhesive. It is bonded to the sheathing and overlaps the drip edge. Some contractors install the builder's felt underlayment to the eave and place the membrane or roll roofing over it, giving double coverage (**6-12**). An easy way to apply it is to remove the protective backing as the material is rolled out over the roof.

UNDERLAYMENT ON LOW-SLOPE ROOFS

Asphalt shingles on low-slope roofs (slope 2:12 to 4:12) require a self-adhered eave-and-flashing membrane be applied from the eave to at least 24 inches inside the interior of the house. Some codes specify 36 inches. Then cover the rest of the sheathing with layers of 36-inch wide, nonperforated asphalt-saturated felt, as shown in **6-13**. Overlap the self-adhered membrane at the eave 19 inches. Each course of felt should also overlap the preceeding course by 19 inches. This gives the entire roof a double coverage of felt. End laps should be 12 inches and at least 5 feet from the end laps in the previous course.

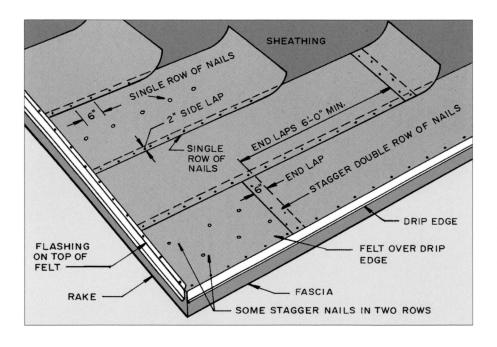

6-8 Lay the underlayment parallel with the eave. It overlaps the drip edge on the eave and goes under the flashing on the rake. Notice the nailing patterns.

6-9 When a roof butts a vertical wall, lap the underlayment up the wall about 4 inches.

6-10 Underlayment is laid over ridges and hips and extends at least 12 inches on each side.

6-11 Self-adhering eave-and-flashing membrane underlayment is used at the eave in areas having considerable snow.

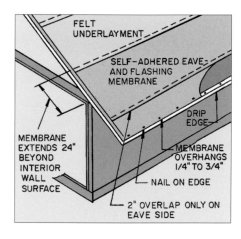

FELT UNDERLAYMENT

SELF-ADHERED EAVE-AND-FLASHING MEMBRANE

DRIP EDGE

MEMBRANE EXTENDS 24" BEYOND INTERIOR WALL SURFACE

MEMBRANE OVERHANGS 1/4" TO 3/4"

NAIL ON EDGE

2" OVERLAP ONLY ON EAVE SIDE

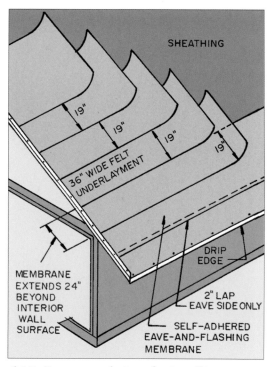

SHEATHING

19"
19"
19"
19"

36" WIDE FELT UNDERLAYMENT

MEMBRANE EXTENDS 24" BEYOND INTERIOR WALL SURFACE

DRIP EDGE

2" LAP EAVE SIDE ONLY

SELF-ADHERED EAVE-AND-FLASHING MEMBRANE

6-12 Some contractors lay the builder's felt to the drip edge and lay a self-adhering membrane or roll roofing over it.

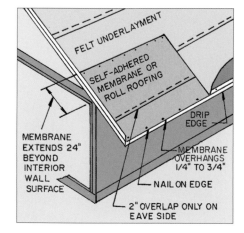

FELT UNDERLAYMENT

SELF-ADHERED MEMBRANE OR ROLL ROOFING

DRIP EDGE

MEMBRANE EXTENDS 24" BEYOND INTERIOR WALL SURFACE

MEMBRANE OVERHANGS 1/4" TO 3/4"

NAIL ON EDGE

2" OVERLAP ONLY ON EAVE SIDE

6-13 Recommendations for installing underlayment and eave flashing for low-slope roofs to be covered with asphalt shingles. This produces a double layer of felt over the roof.

ASPHALT COMPOSITION SHINGLES

The shingles used must have factory-applied, self-sealing adhesive strips. In addition, apply a one-inch spot of asphalt roofing cement under the corner of each tab and press it firmly in place.

ROOFING NAILS & STAPLES

Roof nails should be galvanized steel or aluminum to reduce chances of rusting. The head should be at least ⅜ inch in diameter and for most asphalt shingles 1¼ inches long; it should be long enough to go through the sheathing. They should be installed straight, with the head flush with the surface of the shingle. Underdriven, overdriven, or crooked installation can lead to unsatisfactory long-time service (**6-14**).

Staples can be used instead of nails if they are placed in the same locations as nails. In some high-wind areas they may be prohibited, so, as always, check your local building codes. The staples should have a 1-inch crown with legs that will go at least ¾ inch into the sheathing. They must be driven so that the crown is flat with the surface of the shingle (**6-15**).

6-14 Roofing nails must be set flush with the top of the asphalt shingle and driven straight through the sheathing.

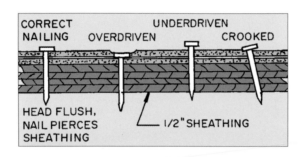

6-15 Power-driven staples must be driven straight through the sheathing and the crown should be flush on top of the asphalt shingle.

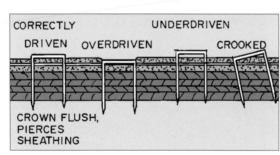

6-16 Roofing cement or some other sealant or caulking is placed in 1-inch spots on the back of the tabs near the butt edge.

SECURING SHINGLES IN AREAS HAVING HIGH WINDS

In areas that occasionally have winds that exceed 60 mph, apply a ¾-inch round dab of a roofing cement, sealant, or a binding caulk on the back of the tabs of the shingles, as shown in **6-16**. This should be applied when the sun has warmed the shingles so that they can be raised enough to apply the cement without damage. Do not apply too much cement at each spot or it may cause blisters or a lumpy appearance. Additional nails can be used.

SHINGLE APPLICATION PATTERNS

There are a number of patterns that can be used for installing asphalt composition shingles. Begin by installing a precut starter course along the eave. These shingles are available in 5- and 7-inch widths. Place the edge with the adhesive strips next to the eave. Allow ½ inch to overhang the drip edge and rake flashing. If these are not used allow the starter shingle to overlap ¾ inch (**6-17**). Starter strips can be made from standard shingles by cutting off the tabs leaving a 7-inch wide strip, as shown in **6-18**. Rather than cutting standard shingles, this strip can be made from rolls of starter strips.

The **6-inch single-column method** is widely used (**6-19**). The first starter-course shingle is cut to a length of 30 inches. After the starter course has been installed, begin the first course of shingles with a full 36-inch-long shingle. Start the second course with a shingle that has 6 inches cut from the end by the rake. It will be 30 inches long. This staggers the spaces between the tabs. Add the additional courses by alternating the beginning shingles using a 36-inch shingle on one course and a 30-inch on the next course along the rake. The rest of the shingles in each course are a full 36 inches long. Notice the spaces between the tabs line up on every other shingle. It is possible that this alignment may contribute to increased wear of the shingle as water is channeled, causing an erosion of the underlying asphalt layer.

Remember to allow the starter course and shingles to overhang the eave and rake, as shown in **6-17**.

Another layout method is the **6-inch, six-course offset method**. This method starts with a full 36-inch shingle at the rake on the starter course. The beginning shingle on each succeeding course is cut 6 inches shorter, with the shingle at the sixth course being only 6 inches long. Then start the seventh course with a full 36-inch shingle and repeat the six offsets.

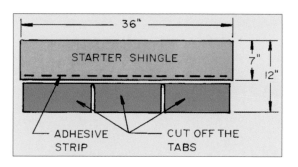

6-17 Begin roofing by installing the starter course along the eave.

6-18 Starter shingles are available precut but can be made from standard shingles.

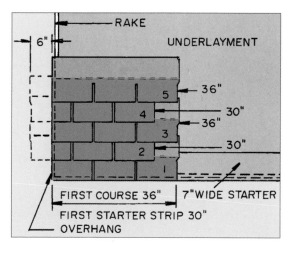

6-19 The 6-inch single-column application pattern begins alternate courses with 36- and 30-inch-long shingles.

ASPHALT COMPOSITION SHINGLES

6-20 The 6-inch six-course offset method reduces the length of each first shingle at the rake by 6 inches until six courses have been installed. The sequence is then repeated.

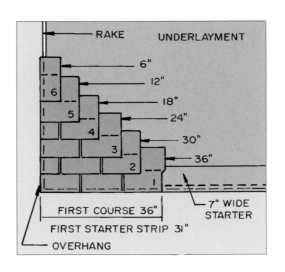

The rest of the courses are filled with full-length shingles (**6-20**). Notice this method also lines up the spaces between the tabs on every other shingle.

As with other methods, remember to allow the starter course and the shingles to overhang the rake and eave, as shown earlier in **6-17**.

Another layout method, the **5-inch step-off method**, has the advantage of staggering the spaces between the tabs, reducing the tendency for a channel of water to flow down these spaces. Begin by installing a full 36-inch shingle at the eave and next to the rake. Cut 5 inches off the next beginning course and install this 31-inch shingle to start the second course. Continue to cut each of the succeeding course shingles 5 inches shorter than the previous shingle for the next seven courses, as shown in **6-21** and **6-22**. Cut the shingle to length, as shown in **6-23**. Measure the length from the right side. This leaves the right side notched, and forming slots between the tabs.

Now fill in the ends of each of these first seven courses with a full shingle (**6-22**). Then lay courses 8 through 12, starting with a shingle 25 inches long, going on to shorten each shingle

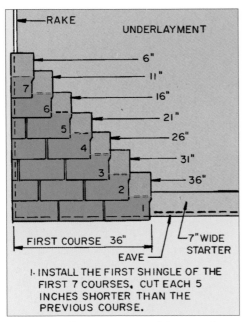

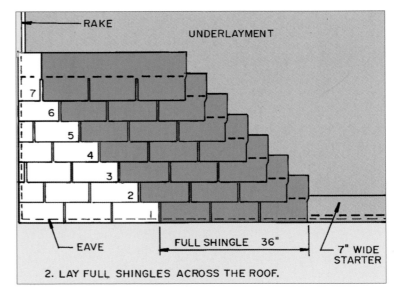

6-21 The first two steps of the 5-inch step-off method that produces good protection by staggering the joints between the butts because they align only every eight courses.

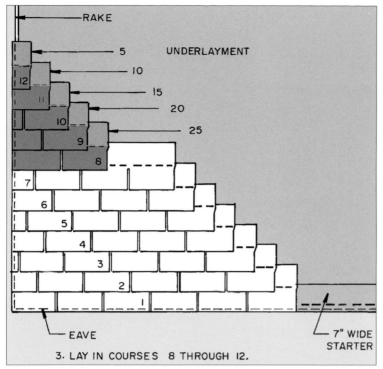

3. LAY IN COURSES 8 THROUGH 12.

6-22 The third step of the 5-inch step-off method that produces alignment only after every eight courses.

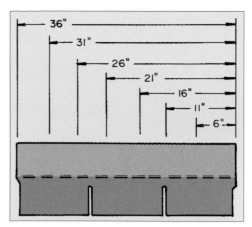

6-23 Measure the length of the shingle from the right side.

above by 5 inches (**6-23**). Then repeat this 12 sheet layout on up the roof.

Again, remember to allow the starter course and shingles to overhang the rake and eave, as shown in **6-17**, on page 79.

INSTALLING THE FIRST COURSE

Since the fascia over which the drip edge was installed is not always perfectly straight, snap a chalk line to locate the top of the edge of the starter strip (**6-24**).

Install the starter strip. It may be a 7-inch wide strip cut from a standard shingle (**6-18**, page 79) or a strip of mineral-surfaced roll roofing. Place the strip with the edge having the self-sealing adhesive beads next to the eave. Let it extend over the drip edge ½ or ¾ inch.

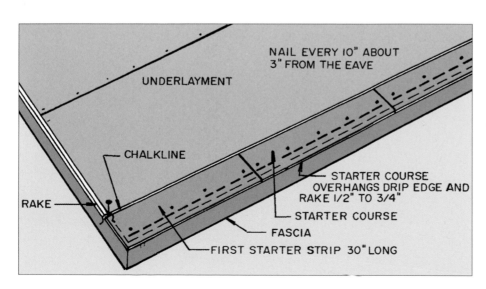

6-24 Stretch a chalk line to get the starter course straight. The fascia may be bowed, and so is not reliable for aligning the shingles.

Nail every 10 inches about 3 inches above the edge of the strip (**6-24**). Cut from the first starter strip a piece equal to the step-off used in the pattern. For example, if using the 6-inch step-off, cut 6 inches off the first starter strip. Lay full-length strips across the roof to the other rake.

Now install the first course shingles. Snap a chalk line 12 inches from the edge of the starter strip. You can lay a shingle on each end of the roof as a measure (**6-25**). Line up the top edge of all the courses above this one from this starting line. Snap more chalk lines every three to five courses above the starting chalk line. If the exposure is to be 5 inches and you locate a line every five courses, the distance would be as shown in **6-26**. As you lay these intervening courses, place the butts on the top course flush with the cutouts in the course below. Some

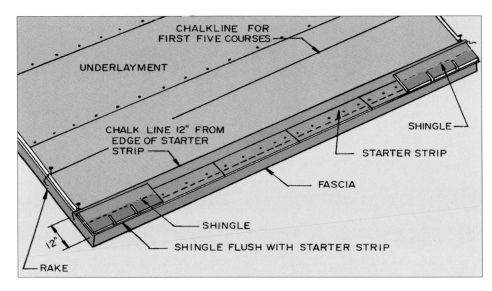

6-25 Locate the line of the first course by measuring up 12 inches from the edge of the starter course. A shingle can be placed at each end of the wall to get the measurement.

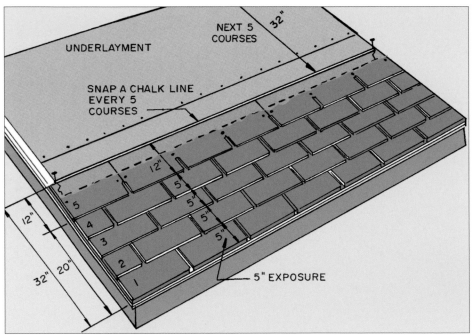

6-26 To keep the shingles absolutely straight, snap a chalk line every three to five rows and work closely to it.

6-27 Shingles are laid in stair-step fashion. The layout depends on the pattern chosen. This example shows the 6-inch single-column method. Lay out the steps in the pattern and shingle across the roof. Then lay out a second stair-step pattern and repeat the installation.

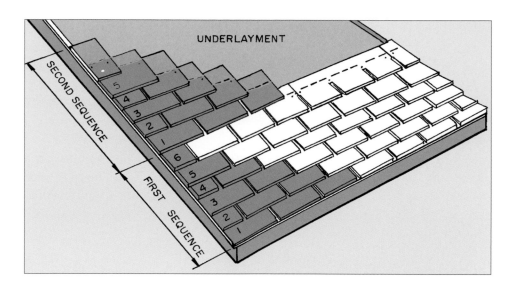

roofers snap a chalk line every 5 inches to locate the top edge of each course.

Lay the shingles in a stair-step fashion as shown earlier in **6-19**, **6-20**, **6-21**, and **6-22**. When you have completed installing the required courses for the first sequence, repeat it forming a second stair-step pattern (**6-27**); continue until the roof is covered.

Remember safety procedures and install 2 x 4 toe boards or a roof bracket.

NAILING PATTERNS

Asphalt shingles used on low- and standard-slope roofs require four nails placed as shown in **6-28**. If staples are used, the same location and placement is used. In areas of high winds, 6 nails are used; staples will likely be prohibited in these areas. Steep-slope roofs, as described below, require roofing cement under the corners of each tab (**6-29**). Check the local building codes.

SHINGLES ON STEEP-SLOPE ROOFS

The application of asphalt shingles on roofs with a slope of 12:12 require one-inch spots of asphalt roofing adhesive under the corners of each tab (**6-29**). Place the spot on the shingle below and press the tab into it.

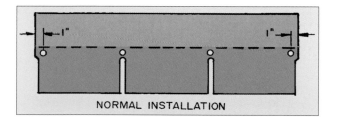

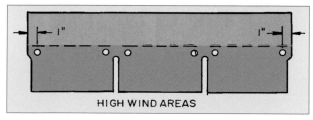

6-28 Recommended nailing pattern for asphalt shingles on low- and standard-slope roofs. Never put a nail through the adhesive strip. Use four nails on each full-length shingle.

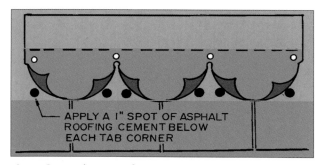

6-29 Steep-slope roofs require a dab of asphalt roofing cement under each tab corner.

CAPPING
THE RIDGE & HIP

Cap shingles for ridges and hips may be cut from standard shingles, as shown in **6-30**, or cut from manufactured cap shingles (**6-31**). If the overlapped part is cut on a slight taper, it will give a neater appearance.

The ridge and hip caps are installed with the same exposure as the shingles. Each is nailed with two nails which are covered by the next shingle (**6-32**). Some contractors put a small spot of asphalt roofing cement under the tab of each cap. Do not put too much or it may cause a bulge. A finished ridge cap is shown in **6-33**.

When the ridge is on a hip roof, the end has a cap made by cutting and folding a shingle as shown in **6-34**.

Ridge vents are widely used, as explained in Chapter 2. Trim back the roof decking a distance specified by the manufacturer of the ridge vent (**6-35**). Then lay and nail the ridge as specified (**6-36**). Finally, nail the ridge-cap shingles over the vent in much the same way as described for a closed ridge (**6-37**).

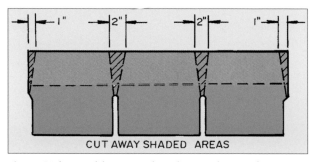

6-30 Ridge and hip cap shingles can be cut from standard shingles.

6-31 Special 36-inch long sections are made to be cut into needed lengths for use as ridge and hip caps.

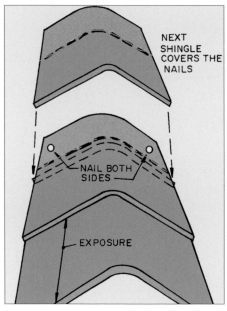

6-32 Install hip and ridge caps with two nails.

6-33 A neatly installed ridge cap.

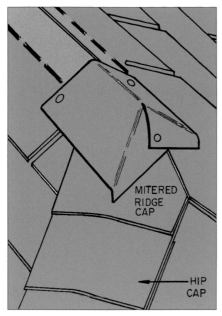

6-34 The ends of the ridge cap on a hip roof are closed with a mitered ridge cap shingle.

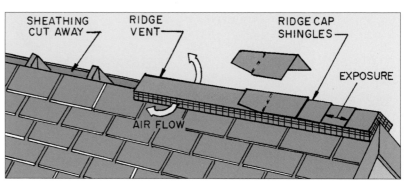

6-35 To install a ridge vent, remove the specified amount of sheathing, nail the vent over the opening, and cover it with asphalt ridge cap shingles.

6-36 Lay out the long section of ridge vent over the ridge opening and nail in place. **Courtesy CertainTeed Corporation and Air Vent, Inc.**

6-37 The asphalt ridge cap shingles are nailed over the ridge vent. **Courtesy Owens Corning**

INSTALLING
INTERLOCKING SHINGLES

Interlocking shingles are especially recommended for areas where high winds are common. There are several shingle designs available; one type is shown in **6-38**. Notice the location of the two nails used in each shingle. The roof sheathing is covered with an approved underlayment, as used with other types of asphalt shingles.

The shingles are locked together by sliding the projections on the tabs into the slots in the shingle below, as shown in **6-39**. Engage one tab into the other slot; push the shingle up until the tab hits the top of the slot.

Begin the installation by installing a 9-inch-wide starter strip. Mineral-surfaced roll roofing is often used. It should overhang the drip edge ½ inch or overhang the fascia ¾ inch if a drip edge is not used (**6-40**). Use a chalk line to line up the starter strip and all courses of shingles.

Now install the first course of shingles. Cut off the tab on all shingles in the first course. Begin installation with the right half of a full shingle (**6-40**). The edge should be flush with the edge of the starter strip.

Start the second course with a full shingle. Put the locking tabs into the cutouts on the bottom edge of the shingles in the first course (**6-41**).

Start the third course with a half shingle using the right-side half (**6-42**); put the locking tab into the cutout on the shingle below. Now repeat this sequence on up the roof. The fourth course starts with a full shingle and the fifth course with a half shingle.

It is important to put several spots of roofing cement along the rake, as shown in **6-43**.

The rake flashing is installed in the same manner as described for three-tab asphalt shingles.

6-38 One type of interlocking shingle available. Check with your supplier or the manufacturers for other designs.

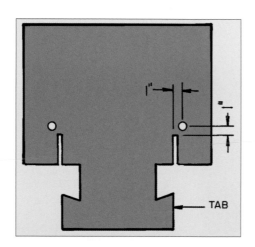

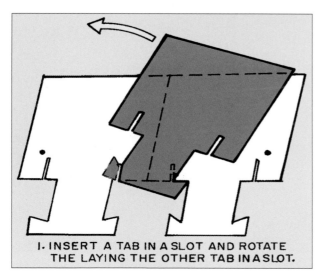

1. INSERT A TAB IN A SLOT AND ROTATE THE LAYING THE OTHER TAB IN A SLOT.

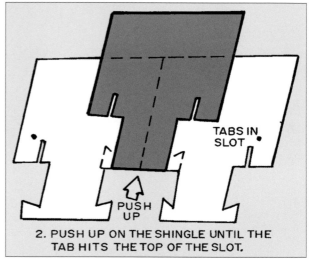

2. PUSH UP ON THE SHINGLE UNTIL THE TAB HITS THE TOP OF THE SLOT.

6-39 One way interlocking shingles are slid together, forming protection against being lifted by high winds.

6-40 Install a starter strip that overhangs the drip edge or fascia. Then lay the first row of shingles. Cut the tab off this first row. Start with a half shingle.

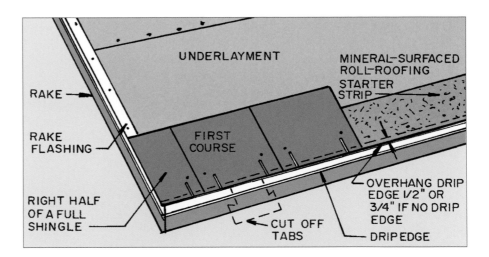

6-41 Start the second course with a full shingle.

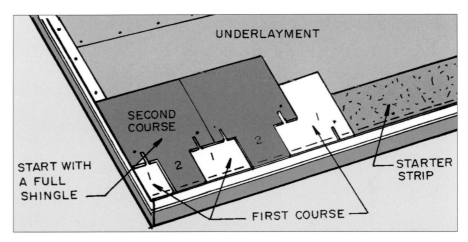

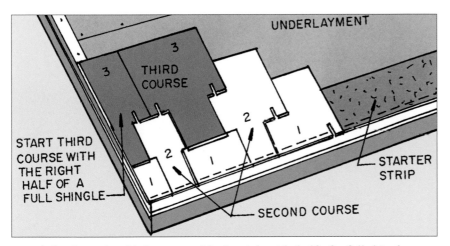

6-42 Start the third course with the right-side half of a full shingle. Repeat courses two and three on up the roof.

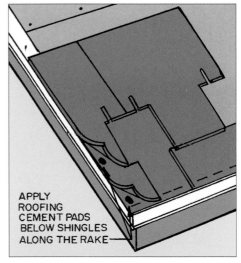

6-43 Apply a dab of roofing cement below each shingle along the rake.

FLASHING VALLEYS

Valleys are often flashed using a woven valley, an open valley, or a closed-cut valley. The woven-valley technique is the one most commonly used, but cannot be used with all shingles.

Each method of valley construction begins with installation of the underlayment over the valley as shown in **6-44**. Then install a metal flashing using one of three methods, open valley, woven valley, or closed-cut valley, as appropriate for the shingles used.

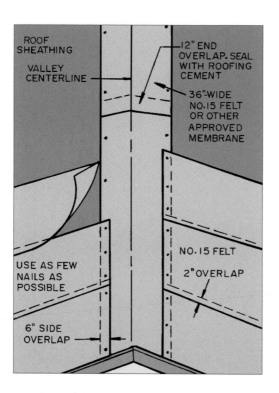

6-44 Valleys have the valley underlayment laid across them and then the roof felt underlayment overlaps it by 6 inches.

WOVEN VALLEY INSTALLATION

This technique is used only with strip shingles, such as the typical three-tab shingle. A finished installation is shown in **6-45**.

Begin by laying the first course over the valley. It should lap over the butting roof at least 12 inches (**6-46**). Next lay the first course along the eave of the butting roof and overlap the first course across the valley and extend on to the adjoining roof. Continue laying successive alternating courses.

Nail the shingles into the valley material but keep at least 6 inches from the centerline of the valley. Place two nails into the very end of the shingle that laps up on the roof.

6-45 The woven valley is easy to install and has a pleasant appearance.

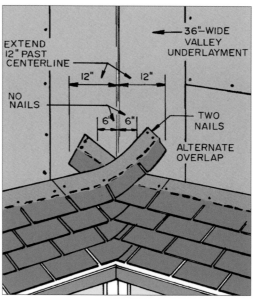

6-46 (Left) A woven valley has alternate courses of shingles overlapping and running over on the adjoining roof.

OPEN VALLEY FLASHING INSTALLATION

Some asphalt shingles have a thickness and multiple layer construction that do not easily conform to the shape of a valley. Manufacturers recommend the use of an open valley with these shingles.

Begin by laying the underlayment over the valley as shown in **6-44**. Then install a 36-inch wide, 50-pound asphalt-saturated felt or a fiberglass-reinforced asphalt-saturated felt designed for use as an underlayment. Some roofers prefer to use mineral-surfaced roll roofing.

The metal flashing, often called W-valleys, should be at least 18 inches wide and is available in 8- and 10-foot lengths. The flashing is secured along the edges to the sheathing by metal clips (**6-47**) spaced every 24 inches. Some roofers pin it on the edges with large-head, rust-resistant roofing nails. The flashing may be aluminum, copper, painted galvanized steel, lead, or zinc.

A typical installation is shown in **6-48**. Overlap the ends of flashing 8 inches and do not nail through both pieces. They need to be separate so they can expand and contract. Set in a sealant.

Strike a chalk line 3 inches on each side of the centerline. Trim the shingles overlapping the flashing to the line and cut the corner on a diagonal. This helps turn the water toward the center of the flashing. Now place a 3-inch-wide strip of asphalt roofing cement on the flashing under the shingle and press them together. Do not nail the shingles.

Some roofers make the open valley wider at the eave because the flow of water is greater as the valley approaches the eave.

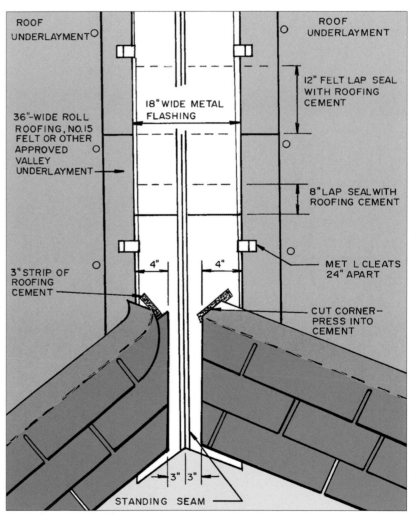

6-47 Metal valley flashing is often secured to the sheathing with metal cleats spaced 24" apart along the edges.

6-48 A typical installation detail of a metal-flashed open valley with asphalt shingles.

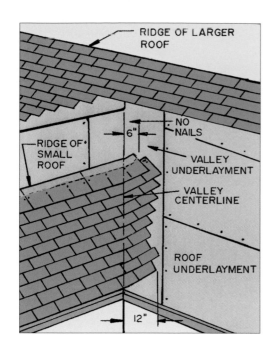

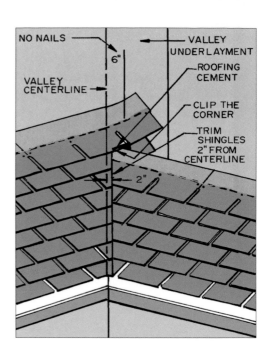

6-49 (Left) To lay a closed-cut valley, run the shingles on the smaller roof over the valley and at least 12 inches over the butting roof.

6-50 (Right) Lay the shingles on the larger roof to the valley. Cut them 2 inches away from the centerline.

CLOSED-CUT VALLEY INSTALLATION

The closed-cut valley flashing technique is used with strip shingles. Flash the valley as shown earlier in **6-44**, on page 88.

Lay the first course of shingles over the valley and onto the butting roof. Lay this course along the eave of the smallest of the butting roofs. Install succeeding courses on this roof in the same manner. Keep all the nails at least 6 inches from the centerline of the valley and put two nails in the end extending over the butting roof (**6-49**). Once the smaller of the roofs is covered, lay shingles on the butting roof overlapping those already installed. Drop a chalk line locating the centerline of the valley; cut these shingles off 2 inches away from the centerline (**6-50**). Also cut about 2 inches on the corner of the top overlapping shingle and seal it with a 2- to 3-inch strip of asphalt roofing cement.

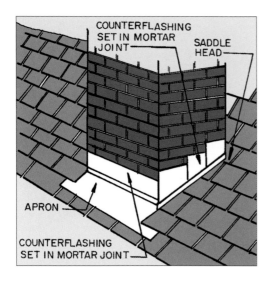

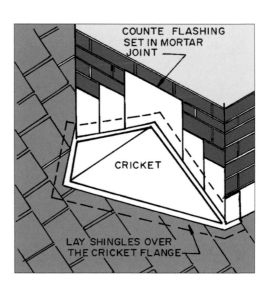

6-51 (Left) Typical flashing around a chimney.

6-52 (Right) An excellent was to flash the back of the chimney is to use a metal cricket.

6-53 When roofs butt a vertical wall, step flashing is used. The siding goes over the flashing.

OTHER FLASHING INSTALLATIONS

The most commonly found areas where flashing is required other than the rake and valleys is where a roof meets a chimney, a pipe penetrates the roof, or the roof meets a vertical wall. These are detailed in Chapter 5, showing the application with wood shingles. Asphalt shingle application is much the same. Some additional information and clarification follow.

A completed flashing installation for a chimney is shown in **6-51** and **6-52**. The counterflashing is set in the mortar between the bricks and laps over the step flashing.

When the roof meets a side wall, such as the side of a dormer (**6-53**), step flashing is used in the same manner as shown for chimneys. A front wall runs parallel with the eave. Flashing for side and front walls is shown in **6-54**.

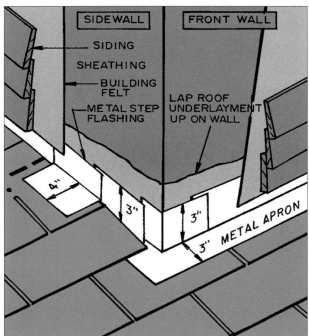

6-54 Step flashing is used when a roof butts a side wall and an apron is used to flash a front wall.

6-55 Steps used to flash a pipe that protrudes through the roof.

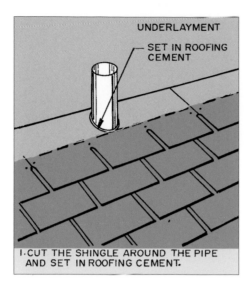

UNDERLAYMENT

SET IN ROOFING CEMENT

1. CUT THE SHINGLE AROUND THE PIPE AND SET IN ROOFING CEMENT.

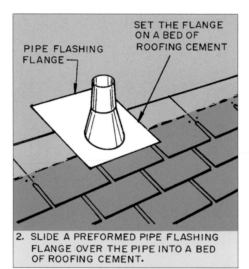

SET THE FLANGE ON A BED OF ROOFING CEMENT

PIPE FLASHING FLANGE

2. SLIDE A PREFORMED PIPE FLASHING FLANGE OVER THE PIPE INTO A BED OF ROOFING CEMENT.

SHINGLES OVER FLANGE ARE SET IN ROOFING CEMENT

3. SET THE SHINGLES AROUND THE FLANGE AND BOND TO IT WITH ROOFING CEMENT.

Pipes piercing the roof are flashed with flashing units manufactured for that purpose. Several different designs are available. The steps are shown in **6-55**. Shingle up to the pipe and cut the shingle to fit over the pipe. Apply asphalt roofing cement around the pipe and press the shingle into it. Then place the flashing unit over the pipe and bond to the shingle below with asphalt roofing cement. Continue to apply shingles up the roof and around the pipe. Cut as needed and set these in asphalt roofing cement. Secure the flashing unit to the pipe as directed by the manufacturer. The unit may be metal or made of a flexible neoprene.

ALGAE DISCOLORATION ON ROOFS

Asphalt roofing becomes discolored due to the growth of algae. This occurs in many parts of North America but is especially common in warm, humid sections. As the algae ages it turns a dark brown and black, causing an unattractive discoloration (**6-56**). It also occurs on wood shingles and shakes, tile, and built-up roofs. It is especially noticeable on light-color roofs.

The common ways to deal with algae are to install zinc strips, install algae-resistant shingles, or wash with a chlorine bleach mixture.

The installation of zinc strips along the ridge of the roof can help keep the roof free of algae infection. It leaches metal ions which inhibit algae growth. These strips are available from several companies.

If new shingles are being installed, then use algae-resistant shingles. Some

6-56 In warm, humid climates algae formations on the roof produce an unsightly appearance.

have cuprous oxide added to the semiceramic color coating. When wet, the copper leaches out inhibiting algae growth. These shingles will inhibit growth for a period of years but not necessarily for the life of the shingle; check the warranty for the protected period. Some algae-resistant shingles have more copper and therefore a longer-term resistance to algae growth.

In areas where algae is common, local contractors are available to spray chemicals on the roof. Some, however, use high-pressure washing systems that tend to dislodge the mineral granules on the surface of the shingles, shortening their life.

There are a number of algae-removing chemicals available. Some recommend using a mixture of chlorine bleach and water; mix 50/50. Another solution recommended includes 1 cup TSP (trisodium phosphate), 1 gallon chlorine bleach, and 5 gallons of water. The bleach can be increased to 2.5 gallons and the water reduced to 2.5 gallons, producing a stronger mixture. The mixture can be varied, depending on the amount of discoloration.

Apply the mixture with a gentle spray or a sponge. Stiff brushes loosen the surface granules and shorten the life of the shingle. After the mixture is on the shingles awhile, wash it clean with water from a hose.

Since the roof is slippery when wet, work from a ladder along the eave. Cover shrubs with plastic drop cloths. As the bleach mixture comes off the roof, it will flow to the ground. When the roof wash is finished and the plastic sheets taken away, spray the shrubs to remove any solution that may have spilled on them.

Remember, handle the bleach and TSP carefully so you do not experience injury. The use of rubber gloves, a dust mask, and some form of eye protection are recommended.

CHAPTER 7

Slate Roofing

Slate is a natural material that when it is processed into shingles, has a variety of textures and colors. The color varies with the quarry from which it is mined.

Slate is a metamorphic rock that began millions of years ago as a sediment deposited under a body of water. Over time it changed under moderate heat and pressure (low-grade metamorphism) into the hard rock that we call slate, which characteristically separates into smooth-faced layers.

Slate has many advantages, including a long life of generally 50 years or more. There are slate roofs in Europe that are hundreds of years old. It produces a roof that enhances the architectural beauty of the home. Since it is fireproof, home insurance rates are lower. It requires little maintenance and, if damaged, individual slate tiles can be replaced. The slate is cleaved along the natural planes of weakness, cut to standard sizes,

and typically comes with holes punched in it for nailing to the sheathing. Finally, a choice of colors and color variations are available (7-1). The dark slate, opposite, blends naturally with the stone chimney and siding.

SLATE SIZES, GRADES & TEXTURES

Slate tiles used in residential construction are generally ³⁄₁₆, ¼ or ⅜ inches thick. Standard thicknesses of up to 2 inches are available. They are cut in 25 to 30 different sizes. Widths from 6 to 14 inches and lengths from 10 to 26 inches are available. The thicker the slate, the heavier it will be per square (amount needed for 100 square feet of exposed roofing surface), as shown in **Table 7-1**, on page 96. A roof built to hold wood shakes, which typically weigh in the range of 400 to 450 pounds per square, will generally

7-1 This slate roof blends well with the brick chimneys and stone exterior wall finish. It has a lighter gray tint than the dark slate, opposite, that is quarried in Vermont. **Courtesy Hilltop Slate, Inc.**

TABLE 7-1 TYPICAL WEIGHTS OF SLATE ROOFING WITH A 3-INCH END LAP

Thickness (inches)	Weight (pounds/square)
3/16	700–800
1/4	900–1,000
3/8	1,300–1,400

Courtesy Vermont Structural Slate Company, Inc.

hold 3/16-inch thick shakes. For slate, however, extra internal bracing is often added or the rafter size is increased to one size larger than that for wood shake roofs. It is helpful to keep in mind, in making a comparison with slate, that wood shakes will absorb moisture during rain or snow and become quite a bit heavier.

If you use **textural** or a **graduated slate** (described below), the roof framing needs to be increased considerably because these are thicker and weigh considerably more than the typical **standard slate**. It could mean roof loads of as much as 1,500 to 5,000 pounds per square. Snow and wind loads also must be taken into consideration.

Slate grades vary depending on the producer and have their own trade names. The properties of slate vary from area to area; so, when a slate is selected, find out the characteristics of the product to be delivered. Some producers have several grades while others have only one. These can be described in detail in writing. If the slate specifications indicate that what is wanted is "commercial standard" roofing slates produced in accordance with the grading standards of the National Slate Association, they will meet predetermined standards for slate produced in a certain area of the country.

Roof slate can be classified by grade according to selected physical characteristics. These include the modulus of rupture across the grain, maximum moisture absorption, and the depth of softening. There are three grades recognized, S_1, S_2, and S_3. Grade S_1 has the best available requirements and has an expected service life of 75 years or more. Grade S_2 has the same modulus of rupture characteristics as S_1, but a higher moisture absorption percentage and a deeper depth of softening; it has an expected service life of 40 to 70 years. Grade S_3 has the same modulus of rupture rating as S_1, but greater moisture absorption and a deeper depth of softening than S_2; it has an expected service life of 20 to 40 years. Additional information can be found in the American Society for Testing and Materials (ASTM) standard *C 406-00*.

Architectural classifications of roofing slate include standard slate, textural slate, graduated slate, and flat slate. **Commercial standard slates** give an average quarry run of 3/16 inch thickness with an allowance for variation above and below this standard thickness. They are graded at the quarry by thickness, length, and width. **Textural slates** are a rough-textured roofing slate with uneven butts. They vary in thickness and size for slates up to 3/8 inch thick. They are delivered in a range of thicknesses and sizes and must be sorted on the job by the roofers. **Graduated slate** is a textural slate (rough texture) of large sizes with variations in thickness, size, and color. The longest, thickest slates are used at the eave. They get thinner and shorter as they approach the ridge.

PREPARING TO INSTALL SLATE ROOFING

Preparing for a roof using 3/16-inch-thick standard slate requires an **underlayment** that is usually 30-pound saturated felt. A typical textural roof will require 30-pound felt. On graduated roofs apply two layers of 30-pound felt under slates up to 3/4 inch thick and 45- or 55-pound felt under the thicker slates. Underlayment is laid as described in Chapter 6.

SHEATHING
FOR SLATE ROOFING

You can use **solid sheathing**, such as plywood, or spaced **wood laths** beneath the slate roofing. Some roofers prefer to use 1 x 6 solid-wood boards instead of plywood. These could be square-edge or tongue-and-groove boards. If tongue-and-groove boards are used, point the tongue toward the ridge (7-2). Use two nails in each board at each rafter. Each end should rest on a rafter. Plywood and OSB (oriented strandboard) panels are installed as described in Chapter 4. Solid-wood sheathing is covered with 30-pound felt.

If **lath boards** are used, they provide an open sheathing. They are generally 1 x 2 or 1 x 3 boards. Some installers prefer this because it allows air to circulate along the inside surface of the slate, drying any moisture that may have accumulated. Some roofers place a layer of 30-pound felt over the lath boards; this protects the interior of the house while the slates are being installed. Other roofers lay felt on the rafters and place the lath boards on top of it. This last method provides an airspace between the bottom of the slate and the felt, which adds to the insulating value. It also helps keep the summer heat from entering the attic, while reducing heat loss somewhat in the winter. However, slate roofing takes much longer to install than most other roofing materials, which could be an important factor in making your selection.

Lath boards do not provide the insulation value given by a roof lined with solid sheathing. In colder climates this is a factor to consider as the roof is designed.

When lath boards are used, solid sheathing is nailed along the eaves (7-3). It is usually set flush with the eave. The slate overhangs 1 inch on the eave and ½ to ¾ inch along the rake. The lath

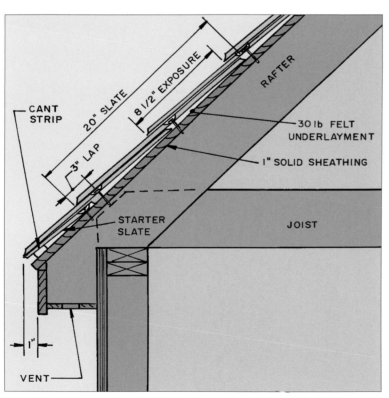

7-2 A typical installation detail for slate over solid sheathing.

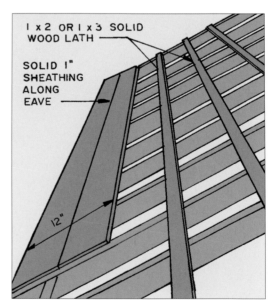

7-3 When the roof is sheathed with wood lath, the eave has 12-inch-wide solid-wood sheathing applied.

TABLE 7-2 RECOMMENDED SLATE EXPOSURES

Length of slate (inches)	Exposure at slope 8" to 20" per foot, 3" lap (inches)
24	10½
22	9½
20	8½
18	7½
16	6½
14	5½
12	4½
10	3½

Courtesy Vermont Structural Slate Company, Inc.

7-4 A typical installation detail for slate roofing installed over spaced lath. This example shows 20-inch-long slate with an 8½-inch exposure.

boards are spaced center to center at a distance equal to the exposure of the slate. Exposures for various sizes of slates are in **Table 7-2**. The slates are overlapped 3 inches and each end rests over a lath board (7-4).

HOLES IN SLATE

Most slates come with two holes punched in them by a machine at the quarry. Large slates require four holes. Typical layouts for holes in slate are shown in **7-5**. The location of the hole varies and this location determines the spacing of the lath in open sheathed roofs. It is possible to specify the spacing of holes when you order the slates or have the manufacturer send several sample slates. Use this information to locate the laths. Once the starting sheathing has been installed, the rest of the laths are placed on a center-to-center spacing equal to the slate exposure.

Slates longer than 20 inches and ¾-inch thick should have four holes. Holes should be punched rather than drilled. Punching produces a countersunk opening around the hole into which the head of the nail will rest. Drilling leaves the surface flat and the head of the nail rests on top of it.

It is necessary to punch holes on the job in slate cut to form **ridge** and **hip caps**. These can be punched with a slate cutter and hole-punching machine (7-6), a hand-held cutter and punch (7-7), or a hammer and slate punch (7-8) or tapped through with a slater's hammer (7-9).

7-5 Hole locations in slate roofing can vary; however, you can specify the location you prefer. Since the location of the holes influences the installation of open sheathing, you need to know what it is before installing the sheathing strips.

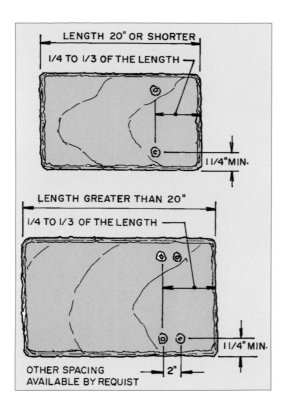

LENGTH 20" OR SHORTER
1/4 TO 1/3 OF THE LENGTH
1 1/4" MIN.

LENGTH GREATER THAN 20"
1/4 TO 1/3 OF THE LENGTH
1 1/4" MIN.
OTHER SPACING AVAILABLE BY REQUIST
2"

7-6 This compressed-air-powered slate cutter and punch is light enough to be carried around the construction site. It is operated on a workbench or a couple of sawhorses. It cuts slate up to ⅜-inch thick and 24 inches long. It provides a fast way to cut slate where many cuts are needed, such as in the tiles along a valley.
Courtesy John Stortz & Son, Inc.

7-7 (Left) This handheld tool will cut slate and punch holes. It will cut slate up to ¼ inch thick.
Courtesy John Stortz & Son, Inc.

7-8 Holes can be punched in slate with a hammer and punch. The hole is punched from the back of the slate with many light taps of the hammer.

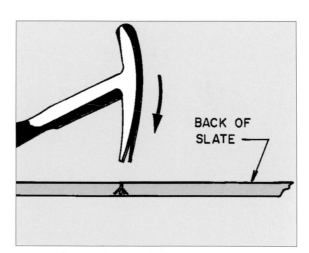

BACK OF SLATE

7-9 Holes can be punched in slate using the pointed end of the slater's hammer. Punch from the back of the slate using many light taps.

TABLE 7-3 NAIL LENGTHS FOR SLATE THICKNESSES

Slate Thickness (inches)	Nail Length (inches)
3/16	1 3/8
1/4	1 1/2
3/8	1 3/4
1/2	2
5/8	2 1/4

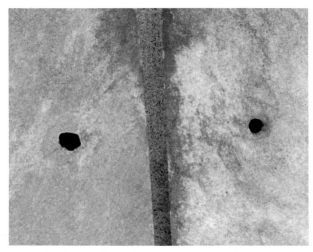

7-10 Back of a slate tile. Holes are punched from the back of the slate.

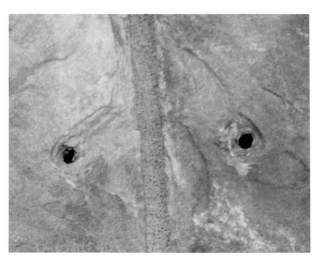

7-11 Top of the slate tile. The breakthrough on the front creates a large round recess into which the head of the nail fits.

When the slate is tapped with short, sharp blows on the back of the slate with the slater's hammer or punch, it will break a small hole through the slate and also break out some of the layers around the hole on top. This forms a conical area for the head of the nail (7-**10** and 7-**11**).

NAILS & NAILING

Since slate roofing will last for a great many years, the nails used should also be expected to serve that long without failure. **Copper nails** are without question a good choice. **Galvanized steel nails** will have a much shorter life. Nails of other materials are available. Research showing their durability should be examined. Never use ordinary galvanized roofing nails as used on asphalt and other types of roofing material. Be certain to use nails designed for slate roofing.

A typical large, flat-headed copper wire slating nail is shown in 7-**12**. It is available in lengths from 1 to 4 inches.

7-12 (Right) The copper slating nail has a large head.

7-13 The slate hammer has a point on one end of the head for punching holes. The other end is used for nailing and has a beveled edge used for trimming slate. It has a claw on the center of the head for pulling nails. **Courtesy John Stortz and Son, Inc.**

Nails are usually driven with a slater's hammer rather than a carpenter's hammer because the slater's hammer has a smaller head and is less likely to strike the slate and break it (**7-13**). The correct nailing technique is shown in **7-14**. The head of the slating nail should just touch the slate. The nail head should touch so lightly that the slate hangs on the nail. Do not drive the nail down tight. Some installers figure the length of nail required by adding twice the slate thickness plus one inch to penetrate the deck. For example, a roof with ⅜-inch-thick slate would use a 1¾-inch nail (⅜ + ⅜ + 1). Typical sizes are given in **Table 7-3**. Remember, longer nails are needed at the hip and ridge.

CUTTING SLATE

Slate can be cut to size on the job using a slate cutter (**7-15**) or a power wet saw that has a diamond blade. The slate cutter is much like a paper cutter. You do not, however, make a single quick slice. Instead you cut the slate with a series of short nibbling strokes, breaking away bits of slate; work your way across slowly. The power saw cuts the slate like a woodworking miter saw; however, it has a diamond blade and plays water at the point of cutting. Always wear eye protection and consider using ear plugs. Another portable tool that cuts slate and punches holes is shown in **7-7**, on page 99. Slate can also be cut using the machine shown in **7-6**, also on page 99.

NAIL DRIVEN TOO FAR. BREAKS OUT BOTTOM OF THE SLATE.

7-14 Slate nails must be driven properly or problems will occur causing roof damage.

NAIL NOT DRIVEN FAR ENOUGH. SLATE OVER IT COULD BREAK.

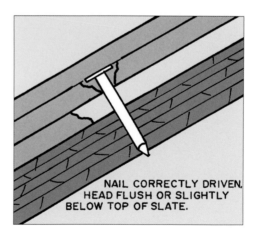

NAIL CORRECTLY DRIVEN, HEAD FLUSH OR SLIGHTLY BELOW TOP OF SLATE.

7-15 (Left) This slate cutter works just like a paper cutter except that the cut is made by using many short shearing cuts rather than one long stroke. It has a small nipping blade for short cuts and can punch two ⅛-inch holes. **Courtesy John Stortz and Son, Inc.**

The slater's hammer has a sharp edge on the shank that can be used to cut slate. It is useful when trimming a piece to fit around a chimney or some other obstruction where a small amount needs to be removed. The slate is placed on the stake that serves to steady it as the cut is made (7-16). The stake has the sharp end driven into a plank or scaffold board and provides a work surface to hold slates as they are cut or have holes tapped into them (7-17).

7-16 This straight slater's stake is used when hand-trimming slate on the roof. It serves as a place to hold the slate as it is trimmed with the slater's hammer.
Courtesy John Stortz and Son, Inc.

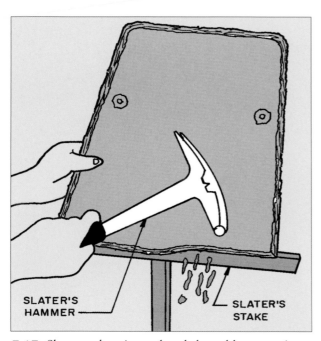

7-17 Slate can be trimmed and shaped by removing a series of small pieces using the slating hammer and stake.

INSTALLING SLATE

When the slate arrives on the job it will be cut in standard lengths. The amount of slate to be exposed to the weather depends on the length. The commonly available slates and the exposure to be used are shown in **Table 7-2**, on page 98. The nail hole is usually 3 inches beyond the edge of the exposure and is covered by the next slate.

The installation of slate roofing is a difficult task and best left to an experienced roofing contractor. It is heavy and the use of power lifts and tractor-operated hydraulic booms are necessary to lift it to the roof (7-18). Scaffolding is also very important along the eave to give a starting working platform and a way to reach the roof (7-19).

Both the underlayment and the slate are very slippery, so extensive use of scaffold brackets is essential. Slate is being installed in a tight, difficult area in 7-20. Some pieces will have to be cut to end each course and holes will have to be punched. The scaffold bracket provides a safe, but minimum, work surface.

7-18 A tractor-operated hydraulic boom is being used to move slate to the roof.
Courtesy Vermont Structural Slate Company, Inc.

7-19 Scaffolding is placed along the eave, providing a working platform. Notice the small lift leaning against the fascia. It will lift several hundred pounds of roofing materials. **Courtesy Vermont Structural Slate Company, Inc.**

7-20 The scaffold bracket provides a level work area from which the slate can be cut, punched and installed. **Courtesy Vermont Structural Slate Company, Inc.**

A finished roof that still has a series of roof brackets that were used to reach the ridge is shown in **7-21**. A tarp has been used to cover the front wall to protect it during the roofing and a scaffold is in place at the eave.

Slate roofs are very slippery and generally used on roofs that are rather steep. Therefore a safety harness and lifeline should be used. Review the safety recommendations in Chapter 2.

INSTALLING SLATE ON SOLID SHEATHING

For residential construction, solid-wood or plywood sheathing is commonly used with slate roofs. Begin by installing the underlayment as illustrated in Chapter 6.

First, you should consider installing a beveled wood cant strip along the eave (**7-22**). This is an optional feature but does help get the starter slate course on the same angle as the rest of the slates. The thickness is the same as the thickness of the slate used.

Second, install the underlayment over the roof, valleys, hips, and ridges. In climates with heavy snows or rains, double the 30-pound builder's felt underlayment at the eave or use a self-adhering underlayment. The self-adhering underlayment is a polyethylene film and a rubberized asphalt coating that has an adhesive on the back. In harsh climates extend this eave underlayment at least 24 inches beyond the exterior wall of the house.

Third, install copper drip edges and rake flashing as shown in **7-23**.

Then begin installing the starter strip (**7-24**). Overlap the drip edge 2 inches. Some roofers place the back of the slate up so that the overhanging edge will do a better job of dripping off the water. Since the joints between slates must be staggered 3 inches from the adjacent courses, use a short starter slate that will locate its joint away from that on the first course. Run chalk lines to

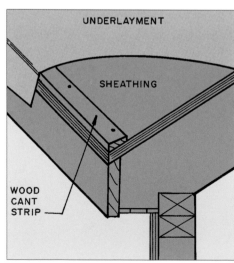

7-22 A wood cant strip is often installed along the eave. It sets the starter course on the same angle as the rest of the courses.

7-23 Install the underlayment over the cant strip and then secure the eave and rake flashing.

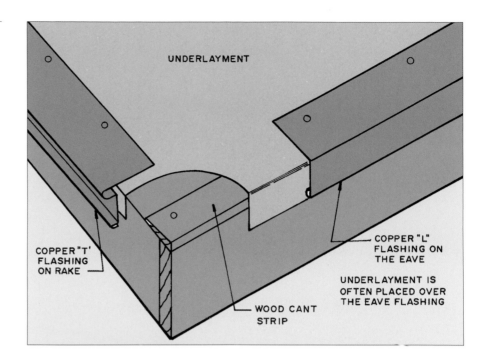

UNDERLAYMENT

COPPER "T" FLASHING ON RAKE

COPPER "L" FLASHING ON THE EAVE

UNDERLAYMENT IS OFTEN PLACED OVER THE EAVE FLASHING

WOOD CANT STRIP

7-24 Install the starter course of slate. Overlay the first course so the joints in each course are 3 inches or more offset.

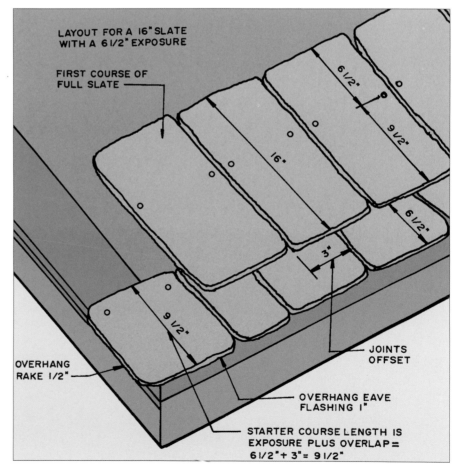

LAYOUT FOR A 16" SLATE WITH A 6 1/2" EXPOSURE

FIRST COURSE OF FULL SLATE

6 1/2"

9 1/2"

16"

6 1/2"

3"

JOINTS OFFSET

9 1/2"

OVERHANG RAKE 1/2"

OVERHANG EAVE FLASHING 1"

STARTER COURSE LENGTH IS EXPOSURE PLUS OVERLAP = 6 1/2" + 3" = 9 1/2"

keep the starter strip and all following courses in line. The courses should overlap the previous one by at least 3 inches to ensure the nails are covered. The slates overhang the eave by 2 inches and the rake 1 inch. It will be necessary to cut some slates to maintain the 3-inch spacing between joints. Measure ahead and plan for the last slate. It should be at least 4 or 5 inches wide; very narrow end pieces will break and are hard to cut, punch a hole in, and nail.

As the second course is laid, begin with a narrower slate to ensure the 3-inch-minimum spacing between joints is maintained.

INSTALLING SLATE ON OPEN SHEATHING

This process of installing slate on open sheathing is much the same as just described for solid sheathing. If the lath boards were properly spaced, as shown earlier in 7-5, the same procedures are used. It is a bit more difficult because the spaces between the lath boards give the roofer no support or place to lay materials. When laying on open sheathing remember the top end of the slate must rest on a lath board.

Begin by laying a solid strip of sheathing along the edge. This should be at least 12 inches wide (refer to 7-5, on page 99). Install the starter course as shown in 7-24, on page 105. Then install the first course. Establish the specified exposure. The slates should overlap the previous course 3 inches, covering the nails. The joints between slates should be offset at least 3 inches. Lay as discussed for slate over solid sheathing. Every nail should hit a lath board. Run chalk lines to keep the courses straight.

INSTALLING RIDGE & HIP CAPS

Typical ridge and hip caps are shown in 7-25. Since the slates are thick, caps become a prominent part of the overall appearance of the roof and the house. There are several ways each of these is laid. The ridge is referred to as the **comb of the roof**, so the slates used to cover the ridge are called **combing slates.** The combing slate should be run with the grain horizontal and the exposure should be approximately uniform.

7-25 Ridge and hip caps form a dominant element of a slate roof. Notice the copper flashing on the chimney and along the edge of the small gable end.

7-26 Install solid sheathing along the ridge if spaced sheathing is used. Lay the last course of roofing slate to the ridge and butt it over the ridge.

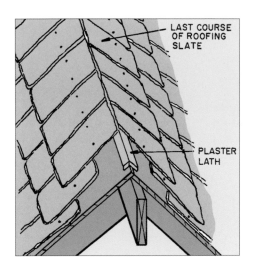

Three ways of laying the combing slates are the **strip-saddle ridge**, the **saddle ridge**, and the **comb ridge**.

If the roof is sheathed with spaced sheathing, install a 12-inch-wide, 1-inch-thick, solid-wood board along each side of the ridge.

The **strip-saddle ridge** has the last top courses of the roof slate butt over the ridge (7-26). Some roofers install plaster lath along the ridge to keep the last courses of roofing slate on the same angle as the sheathing. This helps lay in the combing slate. The combing slates are laid over this and overlap at the ridge. They are laid flat on the last slate and are butted end to end. The combing slate is nailed through the joints between the roof slates below and a large pad of roofing cement is laid where the ends butt. The exposed nails are covered with a silicone caulk and the seam at the ridge is also caulked (7-27).

The **saddle ridge** requires that the top corners of each of the last course of roofing slate be trimmed so that the coursing slate nails go directly into the sheathing (7-28). The combing slates are overlapped, leaving an exposure approximately like that used on the roof; however, the spacing requires that the nails do not hit the roofing slate below. They are installed with two nails that are covered by the overlapping slates. A dab of roofing cement is applied on the end of each combing slate bonding it to the overlapping slate. The overlapping slates at the ridge are caulked.

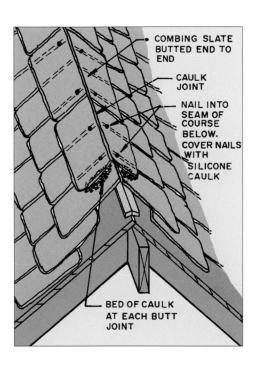

7-27 (Left) Lay the combing slate along the ridge. These strip-saddle combing slates are butted end to end.

7-28 (Right) The saddle-ridge combing slates overlap each other. They are nailed into open spaces made by notching the corners of the last course of roofing slate.

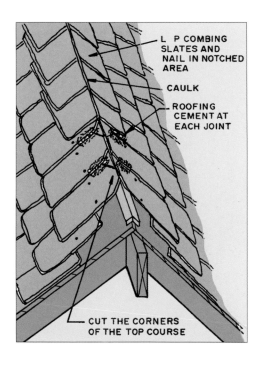

Another strip saddle technique used with slate is called the **comb ridge.** It is installed in the same way as the strip saddle ridge; however, the combing slate on one side extends beyond the other by about ⅛ inch (7-29).

Hips are often covered using the technique described for the saddle ridge. The **saddle hip** requires that two plaster laths or a 3½-inch-wide beveled cant strip be applied along the hip; this makes a smooth transition from the thick roofing slates (7-30). The last course of roof slates butt against the lath or cant strip. The slates are installed using the same exposure as the roof slate. This lines up the ends of each hip slate with the roofing slate. The slates are installed with three nails and a dab of roofing cement is laid over them. Some roofers prefer to omit the plaster lath or beveled strip.

The strip saddle technique can also be used on hips but the laps will not line up with the butts of the roof slate. It is not as weather tight as the saddle hip technique.

Another hip technique is to butt and **miter** the roofing slates along the hip as shown in 7-31. If the mitered slates are chipped like those in 7-32, it is referred to as a **fantail hip.**

FLASHING

Start by reviewing the discussion of basic flashing techniques in Chapters 5 and 6. The techniques for slate roofing are very similar. Since slate roofing lasts for a very long time, the use of copper flashing and copper nails is recommended.

FLASHING A VALLEY

The **open valley** is possibly the most widely used valley-flashing technique (7-33). It is made using 18-inch-wide, 16-ounce copper sheet. It may have a crimp down the center forming a W valley or be unaltered (7-34).

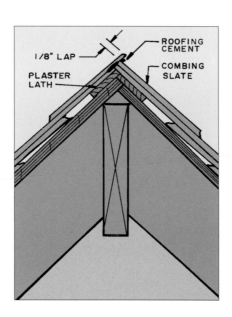

7-29 When the strip-saddle technique is used on the ridge and one combing slate extends beyond the other, it is called a comb ridge.

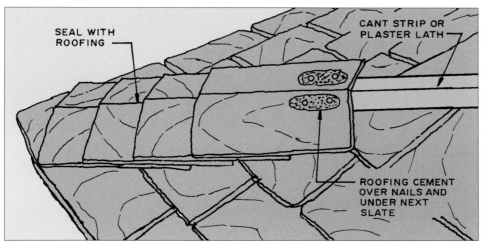

7-30 The use of overlapping slates along the hip forms a watertight seal. It uses the same technique as the saddle ridge.

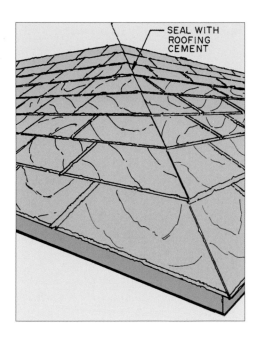

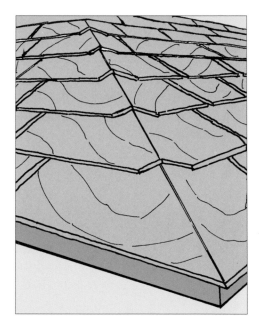

7-31 (Left) A mitered hip has the roofing slates cut to form a joint over the hip. Be certain to seal the joint with roofing cement.

7-32 (Right) The mitered hip can have the corners trimmed, forming a fantail hip.

The W valley has the advantage of breaking the sideways flow of water off the slates.

The valley is covered with underlayment, over which the valley flashing is applied. The best installation uses metal clips to secure the valley flashing to the sheathing. The edge of the valley is rolled, forming a cleat into which the clip is placed (7-34). The clip is nailed to the sheathing with two nails. The rolled edge also helps prevent water from running under the slate.

Overlap the sections of copper valley by 8 inches and bond with a sealant. Do not nail the ends together; they need to be able to expand and contract.

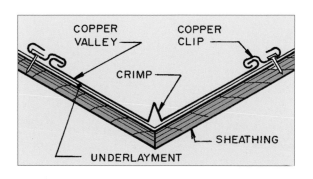

7-33 (Above) This slate roof has a copper-flashed valley. Generally the valleys are tapered so they are wider at the eave than at the top of the valley.

7-34 (Left) The W-valley is widely used. The crimp down the center helps control the flow of water.

A typical open valley installation is shown in 7-35. No nails are permitted to penetrate the copper flashing. The slate is laid 4 inches over the flashing. It is recommended that the slates be laid closer together at the top of the valley with the open valley widening as it approaches the eave. A taper of about ½ inch for every 8 feet of valley is used. For example, an 8-foot-long valley would have the slates extending 4½ inches over the valley at the top and 4 inches at the eave, making the valley 1 inch wider at the eave than at the top.

Run a chalk line locating the edges of the slate over the copper valley. This will establish the taper. Cut the slate to this and punch the needed holes to secure it to the sheathing.

A **closed valley** installation is shown in 7-36. The installation is made in the same manner as the open valley; however, the slates are laid closer to the crimp in the center of the valley. This does cause some problems in nailing the slate over the valley. Avoid driving nails through the copper valley. If you end up with a small slate over the valley you can secure it with a wire as shown in 7-36 or use a wider slate as a substitute for the end slate and the small piece.

Some slate roof installers prefer to actually butt the slates in the center of the valley so that little or no valley flashing is visible. If this is to be done, special flashing techniques along the valley are required.

7-35 Typical installation details for an open copper valley installed with slate roofing.

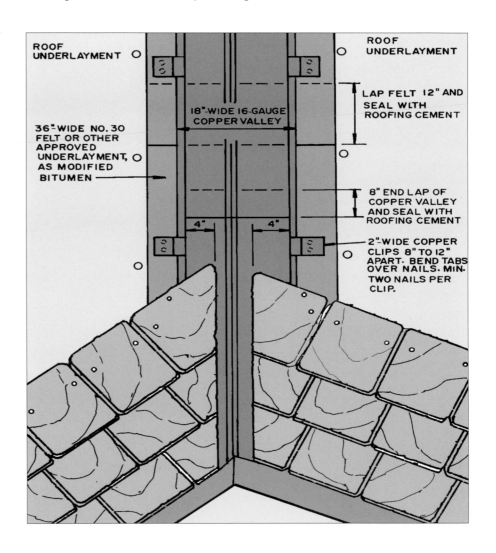

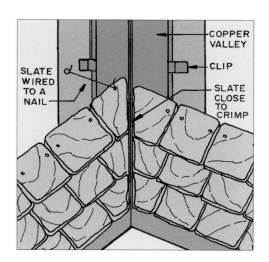

7-36 A typical closed-valley installation using a valley with a crimp down the center.

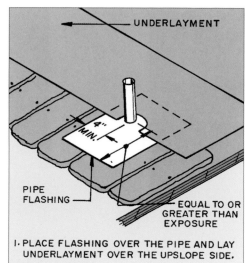

1. PLACE FLASHING OVER THE PIPE AND LAY UNDERLAYMENT OVER THE UPSLOPE SIDE.

FLASHING PIPE PENETRATIONS

Pipes penetrating the roof are flashed as shown in **7-37**. While the slates will not always work out as evenly as shown, the idea is to cover the back of the flashing flange with builder's felt and carefully cut and fit the slates around the pipe. Avoid, when possible, nailing through the flange. Seal the slates to the flange with roofing cement.

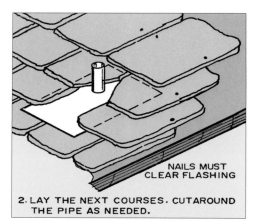

2. LAY THE NEXT COURSES. CUTAROUND THE PIPE AS NEEDED.

7-37 (Above and Left) After installing the pipe flashing, cut and lay the slate courses around the pipe.

FLASHING AT A WALL

When the slate roof butts a wall parallel with the eave, an apron is used to flash the intersection. It has the corner pieces soldered to it. The underlayment is lapped 4 inches up on the wall (**7-38**). When the roof butts a side wall, such as at the side of a dormer, copper step flashing is used, as shown in **7-38**, at right. Refer to Chapter 5 for additional details. The edge of the apron and step flashing are covered with builder's felt and then the siding is used to finish the wall.

7-38 Typical flashing detail when a slate roof meets walls.

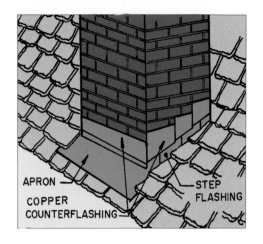

7-39 (Left) The downslope side of a chimney is flashed with an apron and the sides use step flashing.

7-40 (Right) The upslope side of a chimney is best if flashed with a cricket.

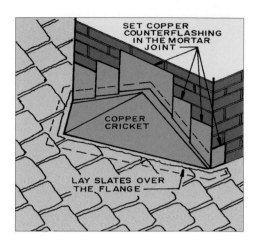

FLASHING A CHIMNEY

A chimney is flashed the same as for wood and asphalt shingles; details are in Chapter 5 and 6. A summary illustration is shown in **7-39**. An apron is used on the down side, step shingles on the sides, and either an apron or cricket on the upside (**7-40**). These are bonded to the chimney and covered with counterflashing that is set into a mortar joint.

REPLACING DAMAGED SLATE

When a slate tile is damaged enough to require replacement, a ripper is used (**7-41**). It is slid under the slate, hooked on the nail and driven back on the handle cutting the nail. Details on roof repairs are in Chapter 3.

7-41 This is a heavy-duty ripper. It is of heavier construction than the standard ripper and is several inches longer. The extra weight reduces the shock to the hand of the roofer as the handle is struck with a mallet as the nail is cut. **Courtesy John Stortz and Son, Inc.**

RIDGE VENTS

Ridge vents are a widely accepted means for venting an attic. They remove hot air in the summer and moisture that may collect in the attic all year long. There are a number of ways ridge vents for slate roofs can be built. One system is illustrated in **7-42**. This system uses a metal ridge cap that is secured to the ridgeboard with a long screw. A metal snap cap covers the screws and seals the ridge. Slate is held in place by a lip on the bottom edge and sealed by the snap cap at the peak (**7-43**). The heavy-gauge-steel unit is designed to last as long as the life of

7-42 This steel ridge vent system provides the needed ventilation at the ridge, completely waterproofs the ridge, and mechanically holds the covering slate. **Courtesy Castle Metal Products**

a typical slate roof, in the range of 75 to 125 years. Since the slates are mechanically held in place, they resist lifting by the wind. The finished installation gives a smooth, pleasing appearance (7-44).

ADDITIONAL INFORMATION

The Slate Book, by Brian Sterns, Alan Sterns, and John Meyer, Vermont. Slate and Copper Services, Inc., P.O. Box 430, Stowe, VT 05672-0430

The Slate Roof Bible, John, Jenkins Publishing Co., P.O. Box 607, Grove City, PA 16127

Slate Roof Quarterly, 50 Rock Road, Ephrata, PA 17522

Slate Roofs, National Slate Association, Vermont Structural Slate Co., Inc., Fair Haven, VT 05743

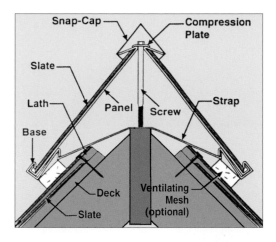

7-43 This section view shows how the ridge vent is installed and how the slate is secured in place. **Courtesy Castle Metal Products**

7-44 The finish ridge vent is a neat, smooth-appearing installation. **Courtesy Castle Metal Products**

CHAPTER 8

Clay & Concrete Tile

lay tile has been used for hundreds of years in countries around the world. The raw material is readily available in many regions. It was used widely in the Middle Eastern countries and especially in China. Ancient Greek and Roman buildings were roofed with clay tile. Settlers of the United States from Europe, particularly Spain, brought the tile roof with them. Clay tile roofs are still widely used around the world.

Today clay tiles are carefully manufactured in a variety of shapes and colors (8-1). They are made from various clays, shale, and other natural earthy materials and fired at high temperatures. Colors can be varied by mixing different types of clays. Clay tiles are also colored by spraying them with a thin creamy layer of clay, which, when fired, forms the color of the exposed surface. The appearance of the tile can be changed somewhat by flowing streams of natural gas into the kiln as the tiles are fired, producing a variegated surface appearance.

Another technique is to coat the surface with a glaze, which, when fired, produces a glassy surface (8-2). The glaze contains metallic pigments, enabling a wide range of brilliant colors to be available.

8-2 This Mediterranean-style home requires the use of clay tile roofing to accurately reflect the architectural style. The tiles have a slight gloss, indicating they have been finished with a glaze. **Courtesy Mr. and Mrs. Conley Williams**

The materials used to make clay tile used in the Unites States must meet the specifications of the American Society of Testing and Materials (ASTM).

Concrete tiles are a more recent development. With the improvement in cements available and the use of additives, lightweight tiles with longterm durability now come in a range of colors, as seen in the photo on the opposite page. They are a mix of cementitious materials, such as portland cement, hydraulic cements, sand, fly ash, pozzolans, and fine aggregates. The materials used must also meet the specifications of ASTM.

8-1 This clay tile roof is the dominant design element, blending harmoniously with the stucco walls and exterior look of the house.

8-3 This concrete roof tile has the color through the entire thickness. It is a flat tile that has a rough surface similar to that found on wood shakes. **Courtesy Vande Hey Raleigh Architectural Roof Tile**

The color of concrete tile can be produced by adding an iron-oxide pigment to the mix that produces color through the entire thickness of the tile (**8-3**). Another technique is to coat the exposed face with a slurry of thin cement mixed with an iron-oxide pigment. This is sometimes used to add a brushed, random second color to the tile; in the photo on page 114, the variegated-surface-finish of the concrete roofing tiles blends with the weathered brick exterior wall and lends to the entire building a balanced, uniform appearance.. Finally, the tiles are sprayed with a clear acrylic sealer. If desired, concrete tiles may be painted with an acrylic medium, such as that used to paint concrete stucco walls; however, this paint does deteriorate over the years and will need to be reapplied occasionally.

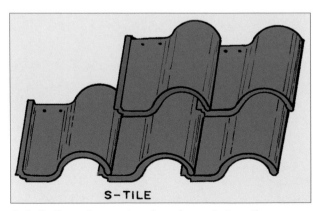

S-TILE

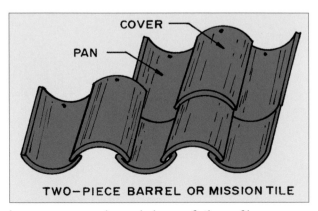

COVER

PAN

TWO-PIECE BARREL OR MISSION TILE

8-4 S-tile and two-piece barrel or mission tile are among the most commonly used clay roof tile profiles.

CLAY & CONCRETE TILE PROFILES

The commonly available clay roofing tiles used on the field of the roof are shown in **8-4** and **8-5** and concrete tiles are in **8-6**. Other profiles are available. Contact a roofing materials dealer for additional possibilities.

Clay and concrete roofing tiles are of the same basic type and installed in the same manner. Both are heavy, so the design of the roof framing must be such that it carries the load. Typical asphalt shingles weigh 235 to 400 pounds per square (100 square feet of exposure). Concrete roof tiles typically weigh 900 to 1200 pounds per square. Lightweight concrete tiles weigh about 550 to 700 pounds per square. A clay tile roof can weigh 800 to 1600 pounds per square. In both cases the specific choice of tile makes a big difference in the weight. Consult the manufacturer for specific weights.

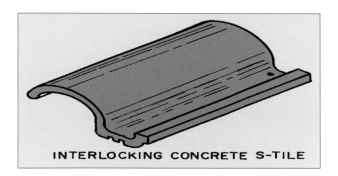

INTERLOCKING CONCRETE S-TILE

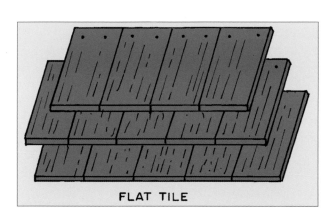

FLAT TILE

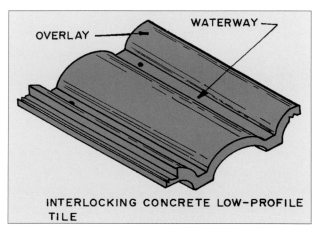

INTERLOCKING CONCRETE LOW-PROFILE TILE

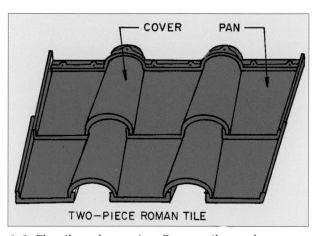

COVER   PAN

TWO-PIECE ROMAN TILE

8-5 Flat tile and two-piece Roman tile are also commonly used clay roof tile profiles. Other designs are available.

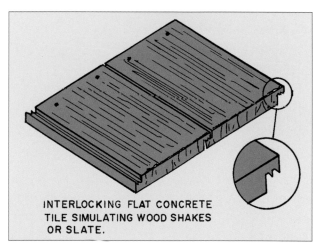

INTERLOCKING FLAT CONCRETE TILE SIMULATING WOOD SHAKES OR SLATE.

8-6 These are the most commonly used concrete roof tile profiles. Other designs are available.

8-7 Cover the entire roof with the underlayment, including the hips, ridge, and valley.

SHEATHING & UNDERLAYMENT

Sheathing must be strong enough to span the distance between rafters and carry the weight of the roof tile. Check the local building codes to verify the approved sheathing materials. Typically, ½-inch or thicker plywood or OSB (oriented strandboard) are used. The panels are installed as discussed in Chapter 4. Leave a ⅟₁₆-inch space between panels to allow for expansion. Solid-wood boards can also be used. Generally 1 x 6 boards will carry the load, but verify this after the tile has been selected.

UNDERLAYMENT FOR TILE ROOFS

The standard underlayment used is 30- or 40-pound builder's felt applied parallel with the eave. It is lapped 3 inches along the horizontal edge and end laps are 6 inches. One layer is applied over the entire roof, including hips, ridges, and valleys (8-7).

Roofs with slopes from 3:12 to just under 4:12 should have two layers of 30-pound builder's felt or one layer of type-90 granular-surfaced asphalt roll roofing.

Roofs with slopes from 4:12 and steeper require only one layer of 30-pound builder's felt. For roofs with slopes of 4:12 to 6:12 it is recommended that the seams between layers be sealed. Slopes above 6:12 do not require sealing.

Roofs with slopes below 3:12 require two layers of sealed underlayment. Tiles on this low a slope are considered decorative, so the underlayment must provide the water repelling function.

In cold climates special provisions must be made along the eave to prevent damage from ice damming. Some roofers install a self-adhering flashing-membrane underlayment along the eave. Other installers double the builder's felt. In any case the extra underlayment should extend at least 24 inches over the interior of the house from the outside wall. See Chapter 4 on Preparing the Roof and Chapter 6 for more details on the roof deck and underlayment.

PREPARING FOR TILE INSTALLATION

The type and design of the tile influence what, if any, further preparation needs to be done before the tile can be laid after the underlayment is down. Some types, such as flat nonlocking shingles, two-piece and one-piece barrel tile, and one-piece S-tile, can be laid directly on the sheathing and nailed to it (**8-8**). These can also be laid over battens. This provides an air space and a means for allowing moisture to drain from under the tiles. Other types have some form of anchor lug and require a series of battens be installed (**8-9**).

Battens are 1×2- or 1×4-inch pressure-treated solid-wood strips. They are spaced to enable the tiles to be nailed into them with a 3-inch head lap (**8-10**). They are in 4-foot lengths and a ½-inch space is left between the butting ends to provide drainage.

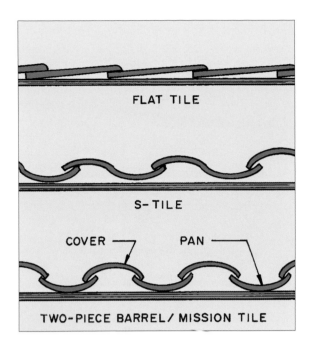

FLAT TILE

S-TILE

COVER — PAN —

TWO-PIECE BARREL / MISSION TILE

8-8 Some types of roof tile can be laid on the sheathing and nailed to it.

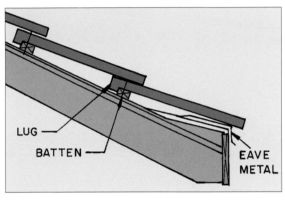

LUG
BATTEN —
EAVE METAL

8-9 Roof tiles with lugs are secured to battens that are installed over the sheathing. **Courtesy Vande Hey-Raleigh Architectural Roof Tiles**

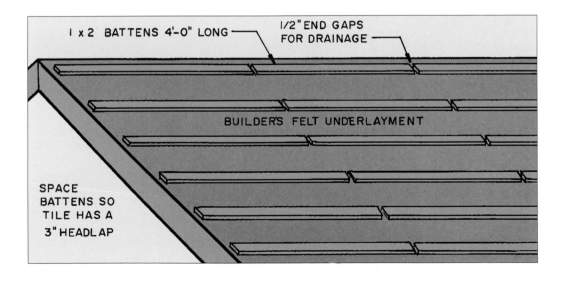

1 x 2 BATTENS 4'-0" LONG

1/2" END GAPS FOR DRAINAGE

BUILDER'S FELT UNDERLAYMENT

SPACE BATTENS SO TILE HAS A 3" HEADLAP

8-10 Battens are installed and spaced to suit the size of the tile to be used.

A system of counterbattens is sometimes used to provide drainage below the tile; such a system provides excellent drainage. The vertical battens should be pressure-treated to resist rot (8-11). The vertical battens can be of ¼-inch or thicker material and nailed to the sheathing with the horizontal battens over them. Battens can be shimmed with a moisture resistant material, such as ¼-inch pressure-treated lath or thick pieces of asphalt shingle (8-12). Another technique is to use pressure-treated wood battens with notches cut every 16 inches, as shown in 8-13.

8-11 Another technique that is used to provide drainage below the tile is to install vertical battens over which the horizontal battens are laid.

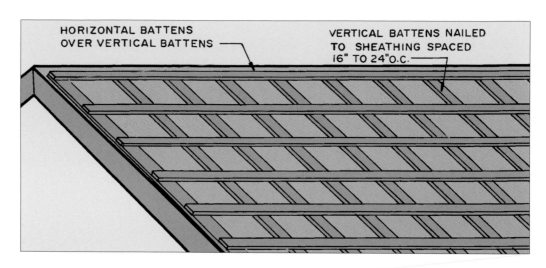

8-12 Shim strips may be used to raise the battens off the sheathing, providing drainage.

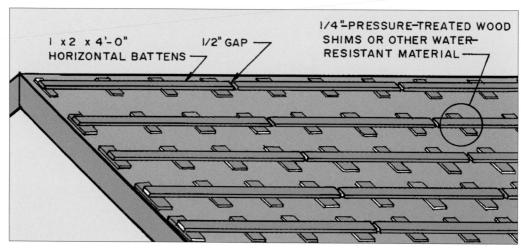

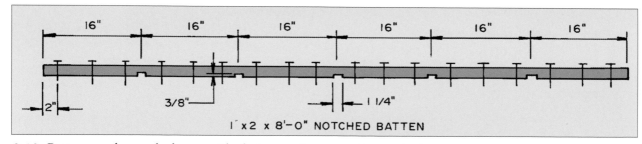

8-13 Battens can be notched to provide drainage. **Courtesy Vande Hey-Raleigh Architectural Roof Tile**

ROOFING MATERIALS & INSTALLATION

8-14 (Right) Battens are placed up to the exposed edge of the metal valley flashing.

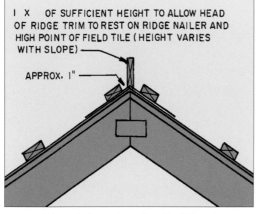

8-15 (Above) A typical detail of framing used at the ridge to carry the ridge tile.
Courtesy Vande Hey-Raleigh Architectural Roof Tile

Battens are placed up to the exposed edges of the valley (8-14).

At the ridge a batten is placed 1 inch from a vertical 1-inch nominal (¾-inch dressed) board that will support the ridge tile (8-15). The height of this vertical board varies but must allow the ridge trim to rest on it. (Look ahead to 8-34 to see a finished installation.)

It is important that the location of the battens be quite precise. Since the tile lug fits over the batten, this automatically lines them up in straight rows across the roof.

If the batten is crooked or has a bow, the butts of the tile will likewise take that shape. Use a chalk line to true up each batten. The spacing between battens is also important. They should be spaced so the required amount overlaps the lower tile and the nail hole hits the center of the next batten (8-16). A typical overlap is 3 inches; the exact spacing will vary depending upon the tile. The nails are covered by the next row of tiles.

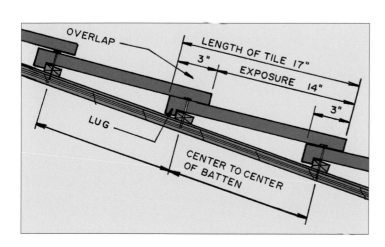

8-16 (Left) The battens must be laid straight and spaced as specified by the manufacturer. This is a generalized example.

CONSIDERING ASSISTANCE & SAFETY

Installing clay and concrete tile is a difficult task and not often tried by the homeowner. The chance for mistakes and accidents is so great it is best to employ a roofing contractor.

One big problem is weight; the tiles are heavy and moving them to the roof requires the use of some type of hoist or tractor with a front end loader (8-17). See Chapter 2 for additional examples of the required equipment.

As the tiles are moved to the roof, they should be stacked in small piles in rows across the entire width. Since they tend to slide and are heavy, the stacks should be kept in short piles: six to eight at a time is ideal. In order to get a good blending of color across the roof, place three or four randomly selected tiles in each pile. Then go back and add three or four more randomly selected tiles to each pile.

If possible, avoid walking on the tile. Not only is it slippery, but it will crack. Replacing a damaged tile is a difficult process.

Wear soft-soled shoes and a harness connected to a lifeline. Review recommendations on personal and common-sense safety in Chapter 2. Keep up to date on current Occupational Safety and Health Administration (OSHA) requirements.

8-17 This hoist will rapidly move the tiles to the roof. The hoist is easy to move around the building to any place that it is needed.
Courtesy Reimann and Georger Corporation

INSTALLING
THE ROOFING TILE

Since there are a number of roofing tile profiles and tile manufacturers, there may be some variations in the installation recommendations. Tile manufactures have detailed installation instructions that are very important to observe. The Roof Tile Institute has installation manuals; see the section on "Additional Information" at the end of this chapter, page 147.

Some roofing tiles require battens to be installed while others are secured directly to the

FRONT

BACK

8-18 This flat concrete roofing tile does not have lugs. Notice the interlocking edges.

sheathing. Spacing and alignment are important. Check how the tile will come out across the roof. You do not want to end up with a very narrow piece at the rake. Some installers measure from the center of the eave to the right and left. Others start on the rake and measure the layout to the other rake.

Use a chalk line to get the starter course straight when installing directly on the sheathing. When battens are used, the battens locate the lines of tile; so they must be straight and spaced correctly.

NAILS

Nails used should be either **ring-shank** or **hot-dipped galvanized**. Smooth nails will tend to lift some over the years.

Most tiles have one prepunched hole with a recessed area around it. This lets the nail head fit below the surface of the tile. If additional holes are needed, they can be drilled with a masonry bit in an electric drill.

Nails should be long enough to penetrate the batten. Tiles laid directly on the sheathing require nails to penetrate ¾ inch into the sheathing. They are driven until the head is flush. Do not drive down hard against the tile; this could break the tile. Also the tile should fit loosely on the nail so that the tile can have some movement. Nails are typically driven with a hammer; however, a pneumatic nail gun can be used, provided the nail depth be adjusted so that it drives to the proper depth.

INSTALLING
FLAT ROOFING TILE

Flat tiles may have lugs that fit over battens or be flat on the back with ribbed, recessed areas (**8-18**). They have a single hole that falls over the batten and are nailed through (refer to **8-16**, on page 121). The hole has an enlarged recessed area around it into which the head of the nail fits, making it flush with the surface.

INSTALLATION ALONG
THE EAVE, RAKE & RIDGE

Construction at the eave to receive flat roofing tile can take several forms, as shown in **8-19**.

A typical installation of rake roofing tile is shown in **8-20**; it uses tile corners nailed to the rake board with two nails. A pad of adhesive in the overlap is recommended in severe climates.

As the rake tiles are laid, they cover one nail and are placed so that they butt against the next field course (**8-21**). Several sealing systems are used.

A typical ridge installation detail is shown in **8-22**. The dimensions of the wood nailer will vary depending on the tile used. The ridge trim is nailed into the ridge nailer. The next nailer overlaps the first and covers the nail (**8-23**).

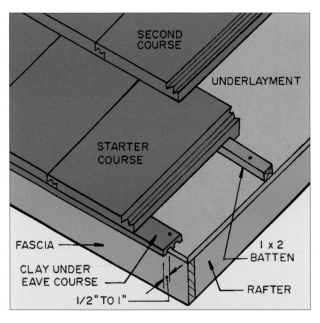

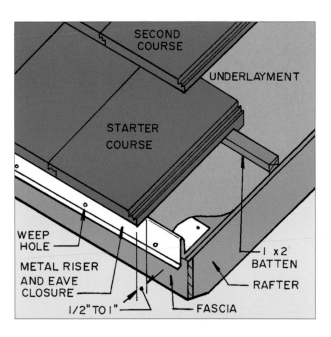

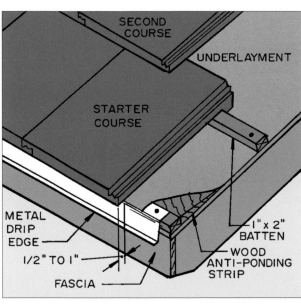

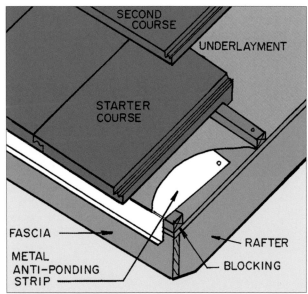

8-19 Four ways to prepare the eave to receive flat tiles.

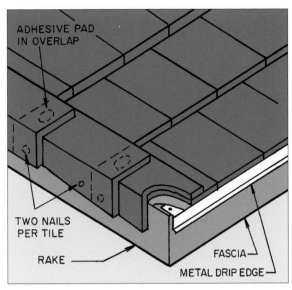

8-20 The rake tiles are held with two nails and overlap each other.

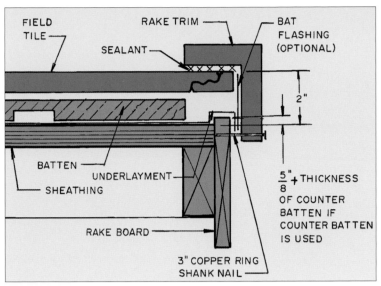

8-21 A section through the rake showing one way to close the edge of the flat field tile and seal it. Other systems are used. **Courtesy Vande Hey-Raleigh Architectural Roof Tile**

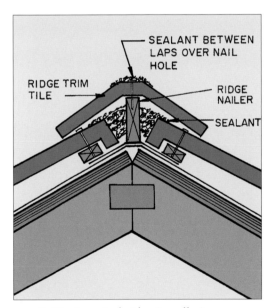

8-22 A typical ridge installation detail showing the use of a wood ridge nailer and the ridge trim tile. **Courtesy Vande Hey-Raleigh Architectural Roof Tile**

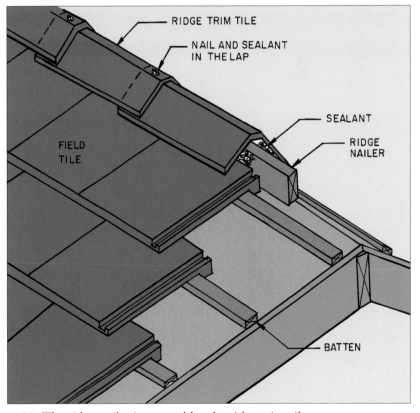

8-23 The ridge nailer is covered by the ridge trim tiles that are overlapped and nailed to the ridge nailer.

CLAY & CONCRETE TILE

INSTALLING TRIM TILE ALONG THE HIP

Installation along the hip rafter is the same as shown for the ridge. A beautiful finished installation is shown in **8-24**. Typically a wood nailer is installed and the hip trim tile is nailed to it (**8-25**). Refer to **8-22** for a typical section detail.

INSTALLING LOW-PROFILE TILE

The interlocking low-profile tiles may have lugs that hook over the batten. A typical concrete tile is shown in **8-26**. It is nailed to the batten strip through two holes located near the top of the tile. A typical detail of the eave is shown in **8-27**.

8-24 This finished concrete tile roof used flat shingles. The shadow lines along the butt of each course add to the overall appearance. Notice the ridge and hip trim tiles. **Courtesy Vande Hey-Raleigh Architectural Roof Tile**

ROOFING MATERIALS & INSTALLATION

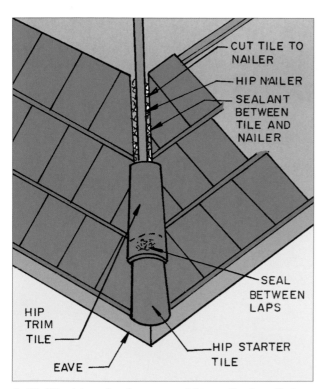

8-25 The hip rafter has a wood hip nailer installed. It is covered with the hip trim tile. **Courtesy Vande Hey-Raleigh Architectural Roof Tile**

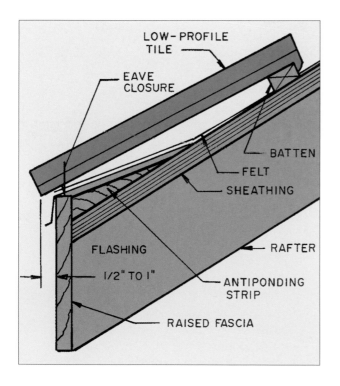

8-26 (Above and top) A concrete low-profile roofing tile. Notice the interlocking edges.

8-27 (Left) One method used to build the eave to receive a low-profile roof tile.

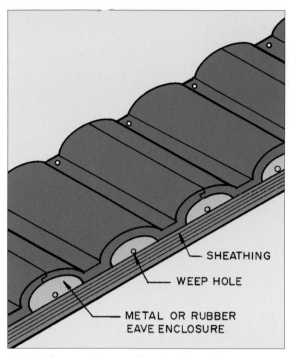

8-28 The openings under the tile at the eave are closed with metal or rubber eave-closure strips.

8-29 Low-profile tile can be nailed to the sheathing. Run chalk lines to keep the courses straight. **Courtesy Monier Lifetile**

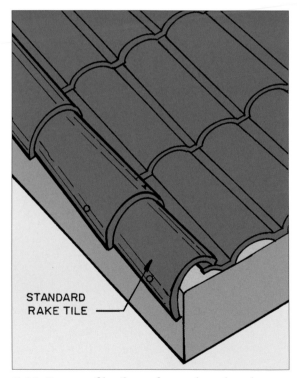

8-30 Low-profile tile roofs can close the edge at the rake with barrel rake tiles.

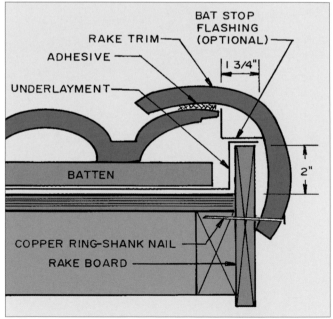

8-31 A section through the rake of a roof finished with low-profile tiles.
Courtesy Vande Hey-Raleigh Architectural Roof Tile

The openings at the eave due to the curvature of the tile are closed with metal, plastic, or rubber eave-closure strips that have a drain hole (**8-28**); they are sometimes called **birdstops**. Low-profile tiles are also laid directly on the sheathing and nailed to it (**8-29**).

One way the roof is closed at the rake is shown in **8-30**. The rake tile overlaps the low-profile field tile and is nailed to the rake board. A section is shown in **8-31**. A finished rake edge for a low-profile tile roof is shown in **8-32**. Notice the valley and ridge trim tile.

The ridge is closed with a ridge trim tile, as shown in **8-33**. They are nailed to a ridge nailer. Sealant is placed over each nail and across the tile, bonding the overlapping tile and sealing out moisture. The roof underlayment is placed over the ridge and a second piece is laid over this to provide extra protection. Eave-closure strips are placed in the concave openings in the field tile. Construction at the hip is much like that described for the ridge. A detail is shown in **8-34**.

8-32 This low-profile tile roof photo shows a rake edge, valley, and ridge trim tile.

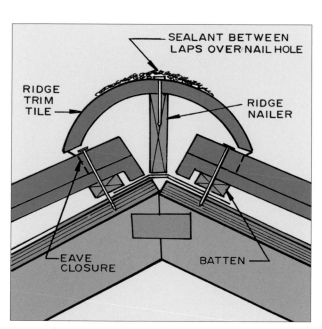

8-33 The ridge is sealed with overlapping ridge trim tile. Notice that the concave opening in the field tile is closed with an eave-closure strip.

Courtesy Vande Hey-Raleigh Architectural Roof Tile

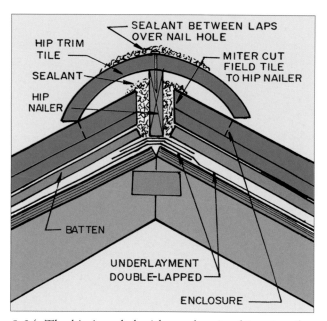

8-34 The hip is sealed with overlapping hip trim tile. Sealant is used to fill the opening between the hip nailer and the edges of the tile.

Courtesy Vande Hey-Raleigh Architectural Roof Tile

CLAY & CONCRETE TILE

The ridge coverage begins with a hip/ridge starter tile, as shown for a hip installation in **8-35**. The hips and ridges are started with a hip/ridge starter tile, as shown in **8-35**.

INSTALLING S-TILE

The S-tile is a medium-profile tile that is made up of a half-round section connected to a flat water course. It is actually a one-piece tile containing a pan and cover, but has a lower profile than the traditional mission or barrel tile (refer to **8-51**, page 137). It is laid in a single thickness and overlaps the adjoining tile 3 inches.

When S-tiles are installed directly to the sheathing, they require separate eave-closure strips (birdstops) to seal the end. They are fastened to the sheathing and the tiles laid over them (**8-36**). When battens are used, an eave-closure strip is used; it raises the end of the tile to provide the slope needed to accommodate the height of the batten (**8-37**). The tile extends ½ to 1 inch beyond the fascia.

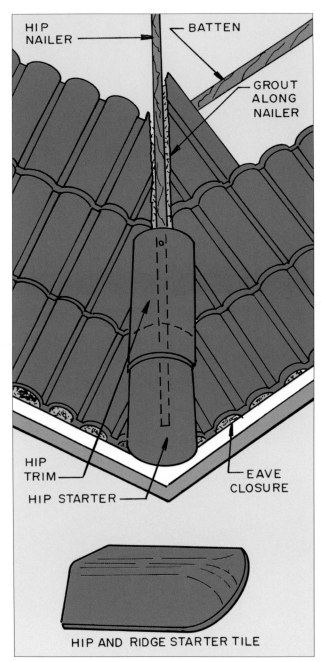

8-35 Hips and ridges are started with a hip/ridge starter tile. **Courtesy Vande Hey-Raleigh Architectural Roof Tile**

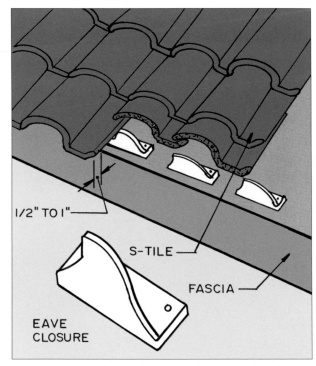

8-36 When S-tiles are laid directly on the sheathing, the openings at the eave are secured with eave-closure strips that are nailed to the sheathing.

8-37 (Right top and bottom) The S-tiles are often raised at the eave to provide the needed slope with metal eave-closure strips.

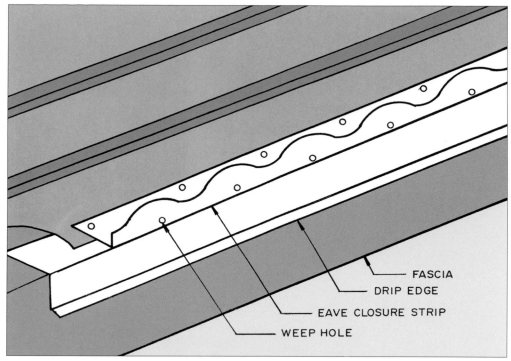

FASCIA
DRIP EDGE
EAVE CLOSURE STRIP
WEEP HOLE

1. INSTALL THE UNDERLAYMENT, BATTENS, DRIP EDGE, AND EAVE-CLOSURE STRIP.

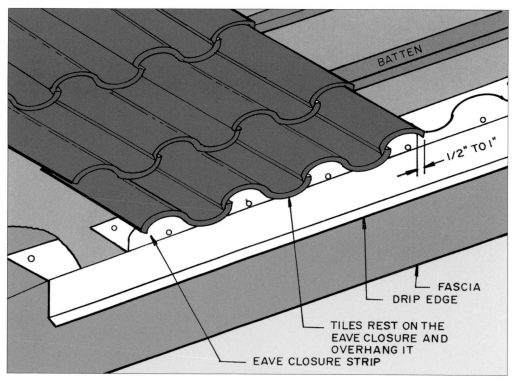

BATTEN

1/2" TO 1"

FASCIA
DRIP EDGE

TILES REST ON THE EAVE CLOSURE AND OVERHANG IT

EAVE CLOSURE STRIP

2. LAY THE S-TILE OVER THE EAVE-CLOSURE STRIP, OVERLAPPING IT ½" TO 1". NAIL TO THE BATTEN.

8-38 These show several situations where the covering of the roof on the rake may require a different solution.

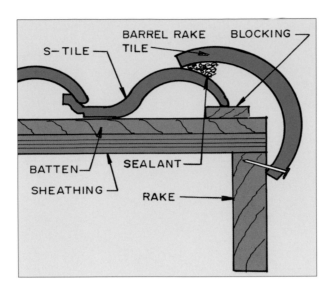

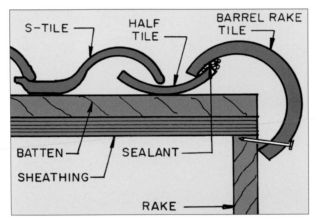

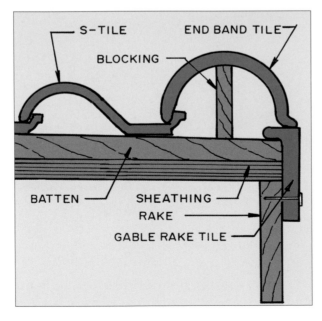

Typical rake installation details when S-tiles are used are shown in the three drawings of **8-38**. Each of the examples uses a special tile made for the particular application.

At the ridge a circular hip-and-ridge tile is used; the installation is similar to that shown earlier in **8-33**, on page 129. The opening caused by the waterway is sealed with a ridge enclosure of some type. Each of the circular tiles is overlapped by the next tile. Sealant is placed over the nail and around the tile to bond the overlapping tile; a complete installation is shown in **8-39**. A distinctive end tile can be used to give a striking appearance as with that shown in **8-40**. Generally a hip-and-ridge starter tile like that in **8-35** is used. (The junction of a ridge trim and hip trim tile is constructed as shown later in **8-47** and **8-48**.)

INSTALLING BARREL OR MISSION TILE

The **barrel tile**, also called **mission tile**, has a high profile and is made up of two pieces, the cover and pan. The high profile provides a dominant feature of the overall architectural appearance (**8-41**).

8-39 This finished installation of a ridge and hip show the overlapping circular hip and ridge tile. They form a dominant part of the overall appearance of the roof.

8-40 This startling ridge cap end tile provides a distinctive ending for the ridge cover.

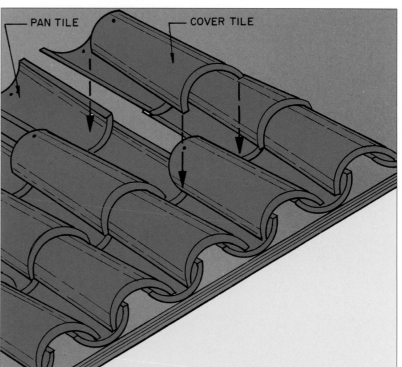

8-41 High-profile barrel tiles dominate the appearance of the house and provide an interesting series of shadow lines.

When barrel tiles are installed, the pan tiles rest on the sheathing and are nailed to it. The cover tiles ride on the pan tiles and they are secured with nails that are long enough to penetrate the wood sheathing (**8-42**). A large dab of sealant is placed over each nail.

Typical barrel-tile construction at the eave begins with the pan tile and clay eave closures set in place (**8-43**). Then the first starter course of cover tile is installed over them (**8-44**). The pan and cover tile are usually extended ½ to 1 inch beyond the fascia.

8-42 Barrel tiles are nailed to the sheathing.

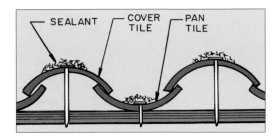

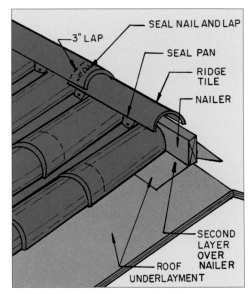

8-43 (Right) Construction at the eave using barrel tile has the pan tile and eave-enclosure strips installed along the eave.

8-44 The starter course of cover tiles are laid over the pan tile and the clay eave-closure strips.

8-45 Typical construction along the rake when barrel roofing tiles are used.

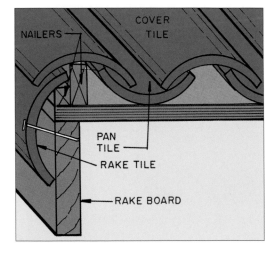

8-46 The ridge is covered with a round ridge tile supported by a wood nailer. The hip rafter is covered the same way. There are several ways to seal the opening between the tiles butting the felt-covered nailer.

Now install the field tile up the roof. Usually a 3-inch lap is used. Use a chalk line to keep the courses straight when laying directly on the sheathing. If battens are used and are accurately placed, they can provide a guide. In any case it is a good idea to snap a chalk line for each course.

A typical construction detail for a rake when barrel tiles are used is shown in **8-45**. It requires blocking to support the last cover tile.

RIDGE & HIP INSTALLATION FOR BARREL TILE

Construction at the ridge and hip rafter is about the same; a typical detail is shown in **8-46**. Blocking is secured at the ridge or hip rafter over which the barrel tile is secured. The blocking is covered with builder's felt that lies down on the sheathing felt several feet. Some type of building code-approved enclosure is used to seal the opening between the roof tiles at the ridge. This could be some type of metal or rubber enclosure. Some installers use a cement mortar to fill this opening. Working on the ridge caps tops off the roof. Since tile can be broken by careless walking and is slippery, this is a dangerous last step. When you start laying the tile on a hip rafter, begin with a prefabricated hip starter tile (refer to **8-35**, on page 130). Use a prefabricated apex trim (wye tile) at the junction of the ridge and hips; this is called the hip apex (**8-47**). Some roofers will install a lead soaker sheet instead. Since lead is soft, it can be worked to fit against the tile and bonded with an adhesive (**8-48**).

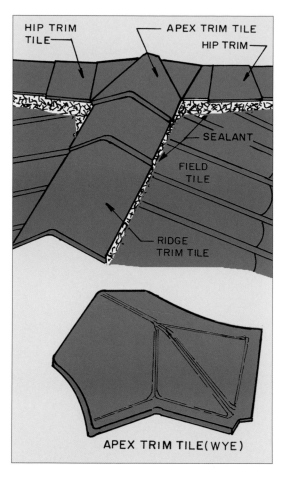

8-47 The intersection between the ridge trim tile and the hip tile is enclosed with a prefabricated apex or wye tile.

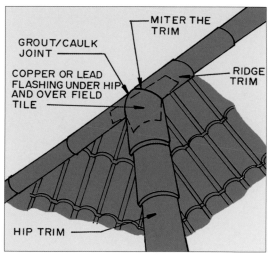

8-48 The intersection of the ridge and hip trim tile at the hip apex can be sealed with lead or copper flashing. The trim tile are laid over it and mitered to fit together.
Courtesy Vande Hey-Raleigh Architectural Roof Tile

Both hip and ridge tile can be installed by nailing, joining with adhesives, or setting in cement mortar.

CUTTING TILES

Concrete and clay tiles can be cut with a diamond blade on a portable circular saw (**8-49**), or a cut-off saw mounted on a stand. If a small area must be removed, such as a corner, and it will be covered, it can be knocked off with a hammer or hatchet. Since this is a dangerous activity, eye protection is mandatory. If much cutting is required, ear protection is also needed because of the noise (**8-50**).

When cutting tile, snap a chalk line locating the line of cut. Cut carefully so the edge is not chipped or corners knocked off.

8-49 Tile can be cut with a diamond blade mounted on a large portable circular saw. Eye, ear, and respiratory protective devices are recommended. Notice the layout of the tiles in preparation for installation. The battens are in place ready for the tile. **Courtesy Vande Hey-Raleigh Architectural Roof Tile**

8-50 Curved tiles are cut with a diamond blade on a circular power saw. This large saw is powered by a gasoline engine.
Courtesy Monier Lifetile

Cutting tile creates a lot of concrete and clay dust. When cutting tile on the roof, you must remove this dust, which is best done with a power blower like that used to blow leaves. Since considerable dust is created, wearing some form of respirator is recommended.

SNOW GUARDS

Snow guards are devices used to prevent massive amounts of snow and accumulated ice from sliding off the roof (**8-51**). These ice and snow packs can accumulate to add considerable weight to the roof load and, if snow guards are used, the roof must be designed to carry these occasional additional loads.

Snow guards are typically used only when there is danger to the passing public. There are many designs, so install where and as recommended by the manufacturer.

8-51 This clay roofing tile has a heavy-colored glaze fired over it. The snow guard is hooked over the top of the tile.

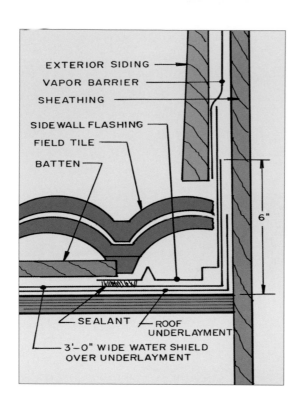

8-52 A flashing detail when low-profile roof tiles meet a vertical wall.
Courtesy Vande Hey-Raleigh Architectural Roof Tile

EXTERIOR SIDING
VAPOR BARRIER
SHEATHING
SIDE WALL FLASHING
FIELD TILE
BATTEN
6"
SEALANT
ROOF UNDERLAYMENT
3'-0" WIDE WATER SHIELD OVER UNDERLAYMENT

FLASHING

The flashing of flat roof tile is much like that used with slate roofing. Review the generalized recommendations in Chapter 7. Flashing low-profile S-tile and high-profile tile employ some special flashing techniques. Flashing is often unfinished 26-gauge galvanized steel; however, prepainted galvanized steel looks better. Since the roof tile will last many years and galvanized steel may fail over the years, consider using copper flashing. It will last the life of the tile; however, it is more expensive than steel.

8-53 This glazed tile roof butts the wall of a dormer on the down slope side. Notice the copper flashing.
Courtesy Mr. and Mrs. Conley Williams

FLASHING
LOW-PROFILE TILE

A frequently occurring flashing job is undertaken when the roof butts against a wall. One way to flash this is shown in **8-52**; the illustration shows it with a batten. The design is the same for tile laid directly on the sheathing. Be certain to wrap the underlayment on the roof up the wall before installing the flashing.

Often the roof butts a wall on an angle. A downslope intersection occurs when it meets a dormer (**8-53**). A detail for flashing this downslope intersection is shown in **8-54**; this illustration shows it butting a brick wall. If the wall has wood siding or stucco, these finish materials will overlap the angle flashing (**8-55**). Stucco requires some special framing to accommodate the thickness of the materials.

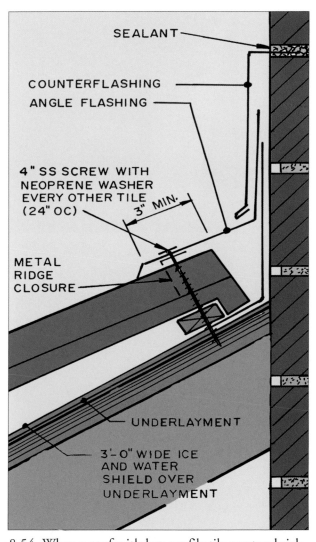

8-54 When a roof with low-profile tile meets a brick wall on an angle, the angle flashing is covered with counterflashing set in a mortar joint. The angle flashing lets the water run on top of the tile.
Courtesy Vande Hey-Raleigh Architectural Roof Tile

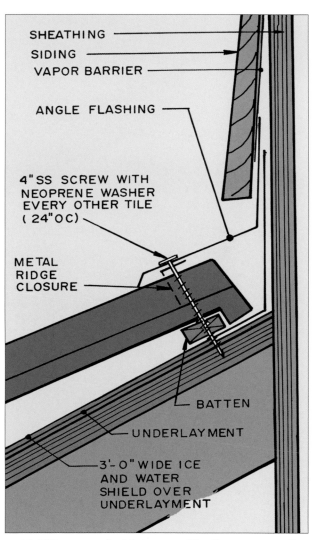

8-55 When a roof with low-profile tile meets a wall with siding, the siding overlaps the angle flashing.

If the roof butts a wall at an angle on the upslope side, it will have an angled backer flashing laid up on the wall that is overlaid with counterflashing. A detail for flashing this upslope intersection is shown in **8-56**. The angle flashing turns the corner and is soldered to the flashing on the side wall (**8-57**).

FLASHING VALLEYS

A typical open valley flashed with copper is shown in **8-58**. A construction detail is shown in **8-59**. If battens are used, they are run to the edge of the exposed flashing area and cut on an angle, lining up with the edge of the underlayment (**8-59**). Review the layout shown in **8-52**, on page 138.

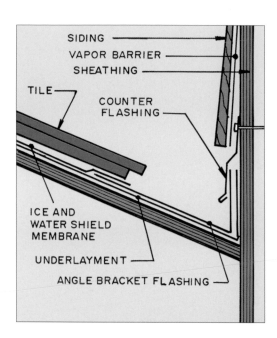

8-56 Typical flashing when a roof butts a wall on the upslope side. The angle backer flashing channels the water to the side where it flows down the side flashing.

8-58 A typical copper-flashed open valley.

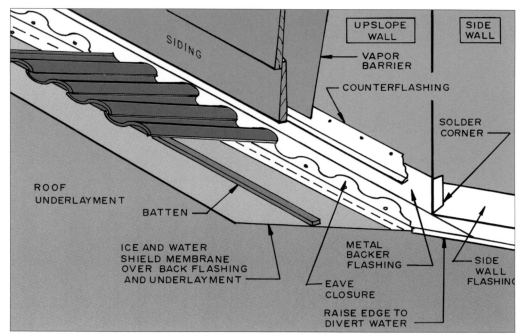

8-57 One way to lay out the backer flashing and tie it into the side-wall flashing.

8-59 A typical
construction
detail for a
valley flashing
with battens.

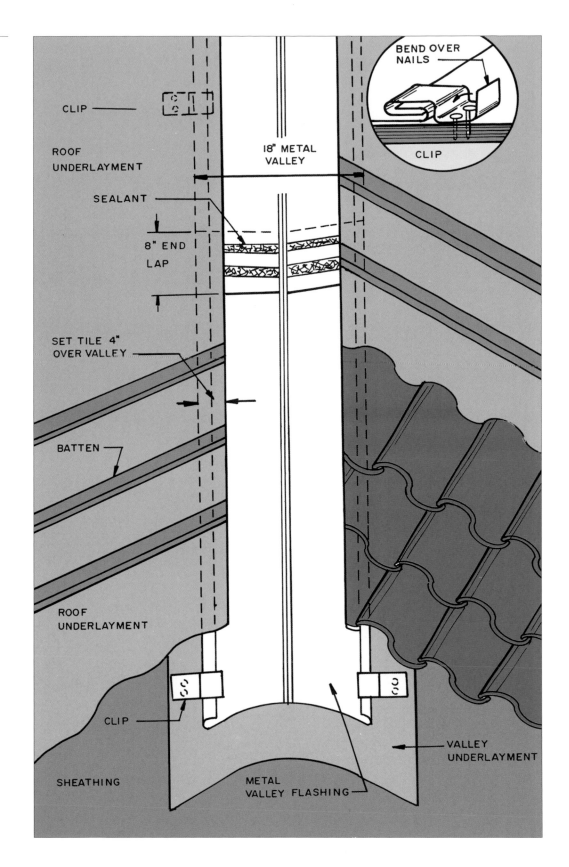

CLIP

ROOF
UNDERLAYMENT

BEND OVER
NAILS

CLIP

18" METAL
VALLEY

SEALANT

8" END
LAP

SET TILE 4"
OVER VALLEY

BATTEN

ROOF
UNDERLAYMENT

CLIP

VALLEY
UNDERLAYMENT

SHEATHING

METAL
VALLEY FLASHING

The valley should be 18 inches wide. The roof underlayment will extend 4 inches over the valley. Some roofers make the exposed width at the eave about twice as wide as at the top of the valley. For example, tile might be placed with the exposed valley 4 inches at top and 8 inches at the eave.

The tiles at the valley have to be cut on an angle parallel with the edge of the metal valley flashing or allowing for the widening of the exposed valley. This can be done by marking and cutting each tile individually or laying the tiles over the metal, striking the line of cut with

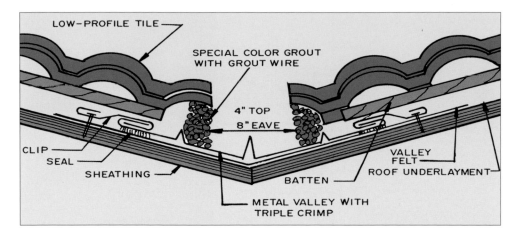

8-60 Detail showing the placement of low-profile tiles along the edges of the valley flashing. **Developed from information from Vande Hey-Raleigh Architectural Roof Tile**

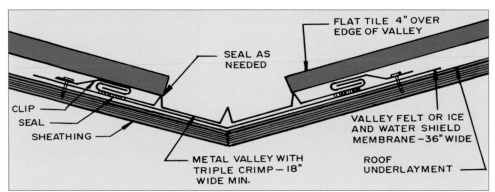

8-61 This detail shows flat roofing tile at the valley. It is laid without battens. If battens are used it will be similar to that in 8-60.

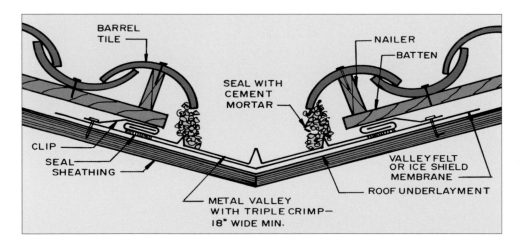

8-62 A section through a valley of a roof using barrel tile. This requires that cement mortar be laid along the edge of the tile.

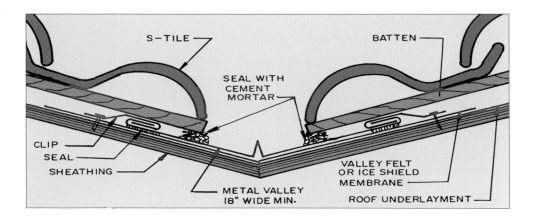

8-63 A section through a valley of a roof with S-tile. This requires that a cement mortar bed be laid along the edge of the tile as it meets the metal valley.

a chalk line and cutting them after installations as shown earlier in **8-49**, on page 136. Be careful not to cut into the flashing.

Following are examples of various ways valley flashing can be used with different types of tile. It is recommended that the instructions provided by the manufacturer be observed.

A section through an open valley with flashing as used with low-profile tile is shown in **8-60**.

The batten is laid over the rolled edge of the flashing. The space created between the tile and flashing is filled with a cement mortar.

A detail for flat roof tile laid without battens is shown in **8-61**. It overlays the edge of the flashing 4 inches. Sections at the valley of barrel tile (**8-62**) and S-tile (**8-63**) are quite similar. Sealing the tile along the flashing with cement mortar finishes the job.

FLASHING CHIMNEYS

For roofs of flat tile, chimneys are flashed in the same way as described for slate roofs in Chapter 7. Other types of clay and concrete tile have curved faces and require a different approach. In **8-64** the upslope roof behind the chimney is flashed with a section of flat pan flashing that is bent up the back of the chimney, forming the backer flashing.

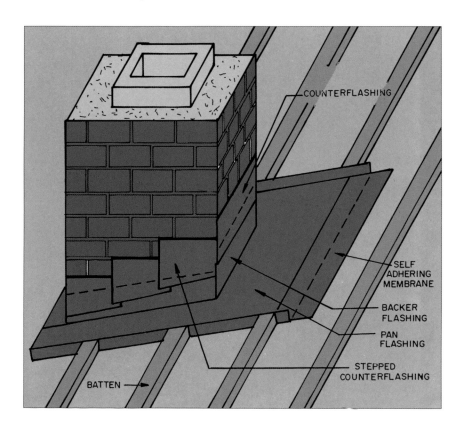

8-64 In this chimney flashing detail, a large pan is used with backer flashing on the upslope side to seal the joint and turn the water. Backer flashing is suitable for chimneys less than 30 inches wide.

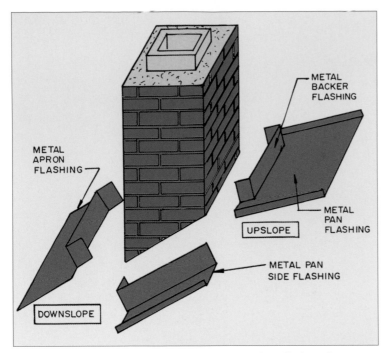

METAL
BACKER
FLASHING

METAL
APRON
FLASHING

UPSLOPE

METAL PAN
FLASHING

METAL PAN
SIDE FLASHING

DOWNSLOPE

8-65 Details showing the flashing units used to flash a chimney.

APRON

FLOW

SIDE PAN
EXTENDS
UNDER
FLASHING

COUNTERFLASHING

8-67 The pan flashing runs down the side of the chimney. It is sealed with counterflashing.

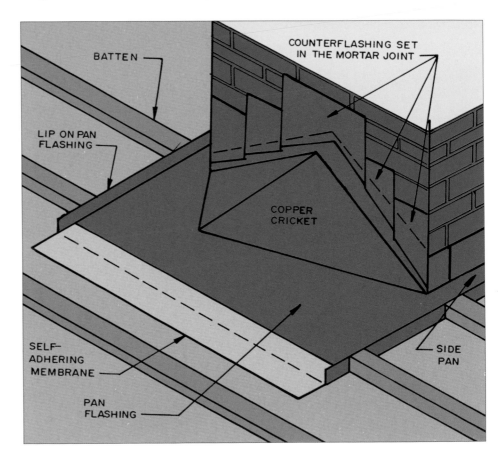

BATTEN

LIP ON PAN
FLASHING

COUNTERFLASHING SET
IN THE MORTAR JOINT

COPPER
CRICKET

SELF-
ADHERING
MEMBRANE

PAN
FLASHING

SIDE
PAN

8-66 Chimneys wider than 30 inches are flashed on the upslope side with a cricket.

The pan is extended down each side to the downslope side. The pan flashing has the edge extend up about 1 inch, channeling the water past the chimney. Counterflashing is laid over the backer flashing and set into a mortar joint. Typical metal flashing pieces that are made in the shop are shown in **8-65**. The joints are soldered. If a chimney is wider than 30 inches, it is recommended that a cricket be installed to divert the water. It is bonded to the brick with an adhesive and sealed with counterflashing set in the mortar joints (**8-66**).

The pan flashing runs down the side of the chimney. One side extends up, under the counterflashing. The lip on the other side is covered by the tile (**8-67**). The water flowing down the side pan flows to the top of the tile below the chimney. A lead sheet is molded to fit the tile and extends under the pan flashing (**8-68**).

FLASHING AROUND PIPES THROUGH TILE ROOFS

There are several techniques that can be used to flash around vent pipes. One technique uses two metal flashing units; the primary unit mounts around the pipe and is fastened to the sheathing. The secondary unit goes over the tile and is exposed to the weather (**8-69**).

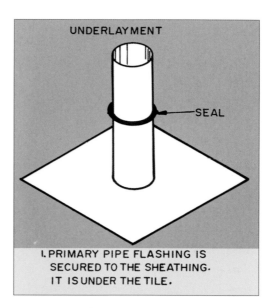

8-69 (Left and bottom) Pipes protruding through tile roofs are flashed with two metal flashing units.

UNDERLAYMENT
SEAL
1. PRIMARY PIPE FLASHING IS SECURED TO THE SHEATHING. IT IS UNDER THE TILE.

SIDE PAN
APRON
LEAD SHEET MOLDED TO TILE

8-68 This construction uses a sheet of lead to direct the flow of water from the side pan onto the field tile on the downslope side of the chimney.

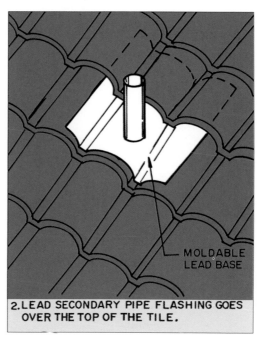

MOLDABLE LEAD BASE
2. LEAD SECONDARY PIPE FLASHING GOES OVER THE TOP OF THE TILE.

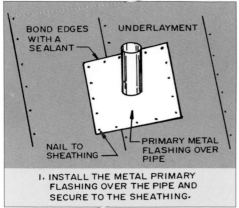

BOND EDGES WITH A SEALANT

UNDERLAYMENT

NAIL TO SHEATHING

PRIMARY METAL FLASHING OVER PIPE

1. INSTALL THE METAL PRIMARY FLASHING OVER THE PIPE AND SECURE TO THE SHEATHING.

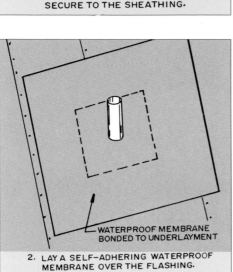

WATERPROOF MEMBRANE BONDED TO UNDERLAYMENT

2. LAY A SELF-ADHERING WATERPROOF MEMBRANE OVER THE FLASHING.

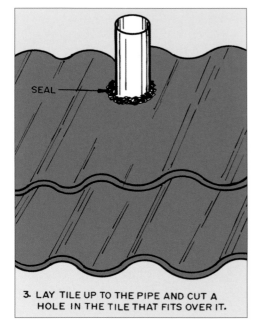

SEAL

3. LAY TILE UP TO THE PIPE AND CUT A HOLE IN THE TILE THAT FITS OVER IT.

SECONDARY LEAD FLASHING

BEND LEAD INTO PIPE

4. LAY THE LEAD SECONDARY FLASHING UNIT OVER THE PIPE. MOLD IT TO FIT THE TILE WITH A RUBBER MALLET. BEND THE END OVER THE END OF THE PIPE.

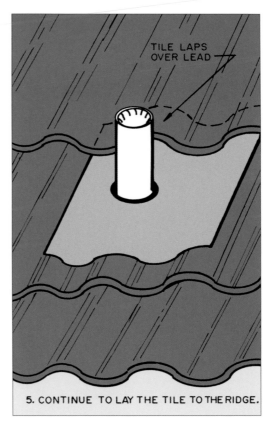

TILE LAPS OVER LEAD

5. CONTINUE TO LAY THE TILE TO THE RIDGE.

8-70 Typical steps used to flash a pipe protruding through a tile roof.

Following are steps for flashing a pipe (**8-70**). Begin by laying a primary metal flashing unit over the vent pipe; bond it to the underlayment with a sealant and nail to the sheathing. Then overlay it with a waterproofing membrane, such as builder's felt, self-adhering membrane, or roll roofing. Cut a hole in the tile that falls over the pipe and set it in place. Seal around the pipe. Then install a lead-pipe-flashing unit over the pipe. Form it to fit the curved surface of the tile by tapping carefully with a rubber mallet. The lead stack around the pipe is bent over the edge of the top of the pipe. Then continue to install the tile on up to the ridge.

8-71 This concrete tile roof provides the dominant feature of the house and connotes a level of quality construction. **Courtesy Columbia Concrete Products, LTD**

IN SUMMARY

The installation techniques shown in this chapter are typical of those in common use; however, individual contractors may have slightly different procedures based on their experiences over the years and the climate conditions in which they build. There are also recognized standards set by organizations, such as the National Roofing Contractors Association and the Roof Tile Institute, and local building codes to observe. Finally, secure the installation manuals from the manufacturer of the tile to be used. Observe not only the techniques recommended but the nails and sealants they recommend.

When all things are considered, roof tiles provide a fire-resistant, weather-resistant roof finish with long-term durability. The choice of color and finish provide the beauty needed to finish the building in a grand style (**8-71**).

ADDITIONAL INFORMATION

Concrete and Clay Roof Tile Design Criteria Manual for Cold and Snow Regions and *Concrete and Clay Roof Tile Installation Manual for Moderate Climate Regions,* Roof Tile Institute, P.O. Box 40337, Eugene, OR 97404

The NRCA Roofing and Waterproofing Manual, National Roofing Contractors Association, O'Hare International Center, 10255 W. Higgins Road, Suite 600, Rosemont, IL 60018

CHAPTER 9

Metal Roofing

Sheet metal of various compositions has been used as a roofing material for several hundred years. Copper was one of the first metals used; later lead proved useful. When you visit historic buildings, look at the roof. Many government buildings and churches had domes covered with small sheets of copper. Within the last one hundred years or so wrought sheet iron was formed into corrugated sheets. While it had a short life, it was less expensive than alternatives. Finally, various coatings were applied to steel, such as hot-dipping the steel into molten zinc, a process referred to as galvanizing the sheet.

Metal roofs have many advantages. They are very durable and lightweight. Currently there are a wide range of profiles and surface finishes. Color of any type can be applied with the quality finishing materials available. Metal roofing in the form of simulated slate, tile, and wood shakes gives this appearance without the weight. Many products are available in long panels that are easy to secure to the wood sheathing with power-driven screws designed to penetrate the panel and secure it to the wood. Installation is rapid, lowering labor costs. Metal roofing is fire-resistant and not affected by UV rays that do damage to most other roofing materials (9-1).

9-1 Metal roofing provides an attractive, durable, weather-resistant, lightweight finish roofing material. There are many designs, colors, and textures available. **Courtesy ATAS International, Inc.**

METAL ROOFING

TABLE 9-1 GALVANIC SERIES OF SELECTED METALS & ALLOYS

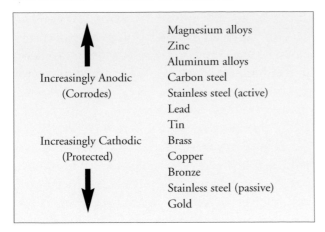

Increasingly Anodic (Corrodes) ↑	Magnesium alloys Zinc Aluminum alloys Carbon steel Stainless steel (active) Lead Tin
Increasingly Cathodic (Protected) ↓	Brass Copper Bronze Stainless steel (passive) Gold

FIRE RESISTANCE

While metal panels themselves do not burn, they do readily conduct heat to the sheathing below. The fire-resistance rating of panels, specified by Underwriters Laboratory tests, is available from the manufacturer. Combustible sheathing can be protected, when necessary, by adding a layer of fire-resistant material over it. Type X gypsum panels are sometimes used. Typically a Class A fire-resistance rating can be achieved. It would actually take considerable fire on the roof to ignite the sheathing below a typical metal roof.

WIND UPLIFT

When selecting a metal roofing material, consider the wind-uplift rating. Some standing-seam roofs have a "Class 90" wind-uplift rating. The design and installation of the eave flashing and rake flashing and the method of securing the roofing to them is critical. Follow the manufacturer's instructions.

PENETRATION

Most metal roofing products are extremely puncture-resistant. Softer metal, such as aluminum, may dent under an extreme hail storm but is seldom penetrated. Metal roofing systems are very durable, having a long life.

ENERGY EFFICIENCY

Since metal roofing readily conducts heat, it will transfer solar heat into the attic, raising the temperature quite high, especially in southern U.S. climates. The sheathing, usually plywood or OSB, provides some insulation. However, additional insulation can be added to the sheathing to reduce heat transfer. When roofing is installed on spaced 1 x 4 boards, heat transfer into the attic is much greater than when solid sheathing is used.

GALVANIC CORROSION

Both ferrous and nonferrous metals are subject to galvanic corrosion. In the presence of an electrolyte (moisture in the atmosphere), dissimilar metals coming into contact will corrode more rapidly than would similar metals. For example, aluminum roofing secured with steel screws produces galvanic action at the point of contact. The presence of moisture in the atmosphere sets up an electrolytic action, causing the aluminum to corrode since it is higher on the table of electrolytic corrosion (**Table 9-1**). The metal highest on the table will be sacrificed. When it is not possible to avoid contact between dissimilar metals, provide some coating or membrane, such as plastic, to keep them from touching. Careful caulking of all joints reduces the chance of moisture reaching the joint and thus reduces the possibility of electrolytic action.

COATINGS

In addition to protecting the metal roofing panels by coating with other metals, such as aluminum, zinc, and lead, they can be painted or covered with a laminate. Laminates are thin plastic films that are adhesively adhered to the panel. The laminates are bonded by heat and pressure and are several times thicker than paint coatings. They resist UV weathering and remain flexible so they do not chip or peel.

PAINT COATINGS

Paint finishes used include polyester resin, silicon-modified polyester resin, and fluoropolymer resin. The polyester finish is inexpensive and has a short life; it will require recoating every few years. The silicon-modified polyester finish has a long life and is possibly the one most widely used. The fluoropolymer-resin finish is the most durable. Paint and roofing manufacturers can provide results of many years' testing.

In all cases use factory-coated material. It is not advisable to install an uncoated steel roofing and try to paint it later; the quality and durability of the finish applied this way is limited. Check to see what warranty is available for chipping, cracking, peeling, fading, and chalking.

METALLIC COATINGS

Steel panels coated with pure zinc are referred to as having been **galvanized.** The coats typically available are termed G-60 and G-90. This refers to the minimum application of the coating. A G-60 coating has 0.60 ounces of zinc per square foot, whereas the G-90 has 0.90 ounces. The thicker the coating, the longer the panel will be protected.

Zinc is a sacrificial coating, meaning it oxidizes as it weathers; this protects the steel panel. When the zinc coating has completely oxidized on some part of the panel it will begin to rust. Obviously, on steep roofs in dry climates, the zinc coating will last much longer than the five to ten years in cold and wet climates. When rust is noted, it is time to clean the panel and apply some type of paint finish over the zinc coating.

ALUMINUM COATINGS

Aluminum coatings on steel are referred to as **aluminized** coatings. The coating used on steel roof panels is termed Type II. Aluminum provides an actual protective coating and is not sacrificed as is the case for zinc. It is a long-lasting, quality coating.

Another metallic coating is a mixture of aluminum and zinc. This aluminum-zinc alloy is referred to as **galvolume.** It tends to heal small scratches through the finish as occurs with a pure zinc coating yet has the permanent protective coating of aluminum. It has a long life and will carry a strong warranty.

THERMAL CONSIDERATIONS

Various metals expand and contract when the temperature changes. This is something that must be considered as metal roofing is installed. Improper installation will cause panels to bow, pull loose from the fasteners, and actually break. The coefficient of expansion of metals used to make roofing varies, with aluminum undergoing the largest degree of change and galvanized steel the smallest.

The amount of expansion and contraction varies not only by the type of metal but also by the color of the finish. A white or light-color roofing will expand less than dark-color roofing. The degree of the expansion movement also varies according to the length of the panel; a longer panel will cause more trouble because it expands more as a single unit. Expansion of several shorter panels will cause fewer problems. A typical panel will have an expansion over its length in the range of ¼ to ½ inch.

Expansion causes the panel to move along the width as well as length. Panels secured to solid sheathing will tend to bow up between fasteners. This pressure on the fasteners can pull them loose or break them. It could enlarge the hole in the sheathing by moving the screw back and forth and eventually make the screw lose its holding power. Aluminum is not as strong as steel and the aluminum may fail at each screw rather than incurring damage to it. This possibility of failure makes installing metal roofing over solid-wood sheathing a problem.

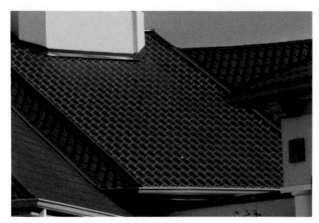

9-2 This aluminum-zinc-alloy-coated steel roofing has been formed to mimic clay tile.

Courtesy Met-Tile, Inc.

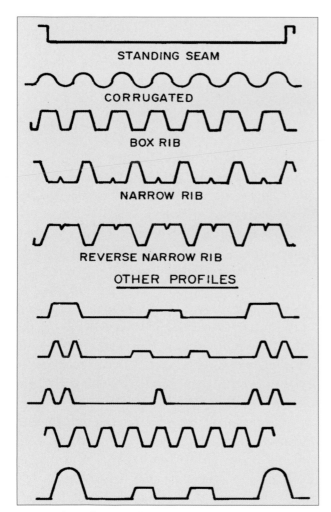

9-3 A few of the profiles available for roll-formed metal roofing panels.

Following are some suggestions that may help. First, use light color roofing; this will reduce expansion and contraction. If installing rolled profile panels with exposed screws, consider using shorter panel lengths; instead of a 30-foot long panel, two 15-foot panels will reduce the total expansion. The overlapping panels when installed on wood sheathing are not screwed together so each can expand independently; however, this technique creates end joints that take time to install and provide a place for a leak to develop.

Standing-seam panels are secured to the sheathing with metal cleats, allowing long panels to expand and contract. A one-piece cleat will allow some expansion; a two-piece cleat will allow greater thermal expansion.

TYPES OF METAL ROOFING

Metal roofing is available in panels that have been roll-formed into a number of patterns, including flat panels with standing seams and corrugated roll-formed panels. It is also available formed into simulated slate and clay tile (**9-2**) as well as wood shingles and shakes, as seen in the chapter opening photo, on page 148, where the metal is colored to appear as weathered wood. The metal units are much lighter and easier to install than actual wood shingles, shakes, clay, or slate roofing.

PANEL PROFILES

Various roll-formed panel profiles are shown in **9-3**. These are formed in a rolling mill where large machines form flat metal panels into the desired shape as they pass through the machine. The mill will manufacture these to the length needed to span distances on the roof; this eliminates the need for end joints. Typically, 36 feet is the maximum length made because anything longer is difficult to move by truck.

While panels already roll-formed are available, the contractor can use a portable roll-forming machine that is moved to the site (**9-4**). The panel material can be cut to length on the site and then the profile rolled (**9-5**). Each piece needed may be made to fit exactly where it is needed. A finished installation is shown in **9-6**.

Roll-formed panels have profiles that can be a series of curves, straight lines, V shapes or some combination of these. The design chosen influences the appearance of the finished roof and the method of joining the panels along the edge.

9-6 These roof panels were site-formed and have a 2-inch standing seam. **Courtesy Knudson Manufacturing, Inc.**

9-4 This portable, standing-seam-panel manufacturing system can be moved to the job site and forms the panels to the width and length required, as below in 9-5. **Courtesy Knudson Manufacturing, Inc.**

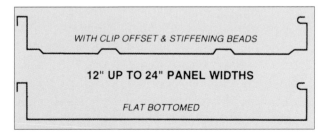

9-5 The panels formed by the system in 9-4 can have a flat or ribbed bottom, in widths from 12 to 24 inches. **Courtesy Knudson Manufacturing, Inc.**

The folds in the roll-formed panels add rigidity and therefore also serve a structural purpose. They produce a repetitive pattern across the roof, giving an appearance not available with other roofing materials. Some patterns give a strong line from the ridge to the eave. Others give a shallow rhythm flowing across the roof.

Consult manufacturers' catalogs to view the wide range of designs available.

TYPES OF PANELS

Metal roofing panels are available in two basic types: **architectural** and **structural**.

Architectural panels must be supported by a substructure, such as plywood sheathing. The minimum roof slope for architectural panels is 3:12. Some form of underlayment, such as builder's felt, is installed on the sheathing before the panels are set in place. These are used on steep slope roofs.

9-7 These granular-coated steel shingles resemble clay S-tile. They are a lightweight, fire-resistant roof finish.
Courtesy Dura-Loc Roofing Systems

Structural metal roofing panels have load-carrying capabilities. They can span distances between supports, so solid sheathing is not required. The span and slope depend on the thickness of the metal and the profile. They generally can be used on roofs with a slope as low as ½:12. The manufacturer will provide span data, but spans up to five feet are available. Any expected loads, such as heavy snow and ice, are another consideration when establishing the unsupported span.

METALS

Commonly used metals include copper, stainless steel, other steels, such as weathering, aluminized, galvanized, and granular-coated steel, as well as tern metal and aluminum.

Copper is the most impressive of the roofing metals. It lasts for hundreds of years. As it ages it develops a brownish-green patina. This patina actually forms a protective layer over the copper, protecting it from corrosion. Since this process takes many years, some try to speed it up with various chemical treatments.

Drainage from a copper roof will produce a stain on painted fascias, walls, and soffits. Since it is low on the galvanic series it resists corrosion but will corrode most other metals.

Stainless steel is a steel alloy containing alloying elements, such as manganese, chromium, nickel, and molybdenum. The higher the chromium content the more resistant the roof panel will be to corrosion. Chromium content will range from 11 to 25 percent. The most commonly used stainless steel for roofing is termed 18/8, which means it has 18 percent chromium and 8 percent nickel. Nickle increases strength and toughness. Chromium gives stainless steel its corrosion resistance qualities and produces a thin, hard, invisible film over the surface, which inhibits corrosion.

Stainless steel has a rolled semibright finish or a rough rolled finish that is not as bright and reflective. It can sometimes develop tiny pits, which may after many years cause a leak. This is aggravated when exposed to the salt in the moisture in coastal areas. Stainless steel can be bonded to other metals without danger from galvanic action.

Weathering steel is a special product that is designed to rust when exposed to the weather. As it rusts it develops a thick layer of oxides, protecting it from additional corrosion. Weathering steel made by United States Steel Corporation is sold under the trade name Cor-Ten. The steel sheets are much thicker than are other architectural roofing panels. It weighs about the same as asphalt shingles. It does produce a rusty runoff down the exterior walls of the building so this must be considered if it is to be used.

Aluminized steel roofing is coated with a layer of pure aluminum. This provides all the characteristics of an aluminum roof panel. This is referred to as aluminized steel. Another type uses an aluminum-zinc alloy coating; those made by Bethlehem Steel Corporation are available under the trade name Galvalume. This product works well in coastal areas where the salt in the air would damage many other materials.

Galvanized steel has a hot-dipped coating of liquid zinc deposited on the surface. The zinc coating has the function of protecting the steel from galvanic action. It is sacrificed protecting the steel. Once it is gone the steel is vulnerable. Galvanized steel roofing is protected with some kind of finish coat, such as paint; both sides of the panel should be painted.

Granular-coated steel roofing uses a Galvalume steel base sheet. Galvalume steel, as mentioned earlier, is coated with an aluminum-zinc alloy. This has UV-resistant acrylic added. The surface is finished with blended or variegated ceramic coatings. Galvalume steel is also available with smooth finishes in copper or zinc. It is formed to the profile of clay S-tile, wood shakes, and slate (9-7). They are considerably lighter than clay tile, concrete tile, slate, or wood shakes yet have the granular coating, giving a realistic appearance and fire resistance.

Tern metal roofing is composed of a mild steel sheet coated with an alloy of lead on tin. It has been in use for many years in Europe and is an excellent product. The tern plating resists weathering but must be primed with red iron oxide in linseed oil and repainted periodically (9-8).

9-8 The tern roofing on the left has been primed and painted. That on the right is ready for painting.
Courtesy Murton Roofing of South Carolina, Inc. and Don Wheeler, Photographer

9-9 This tern metal roof has been in place for more than 75 years. **Courtesy Follansbee Steel**

Stainless steel that has been tern coated will not rust and the coating will weather to a natural dark gray. Tern metal roofing has a long life, as shown by the classic house in **9-9**; this roof has been in place more than 75 years.

Aluminum roofing materials are an alloy of pure aluminum that is weak and soft and other metals, such as manganese. This material resists corrosion, including salt damage in coastal areas. If used uncoated, it starts out with a rather bright finish but over the years weathers to a gray color as an oxide coating builds. This coating also slows corrosion. Aluminum roofing can be painted, anodized, or protected with other coatings (**9-10**).

INSTALLING
METAL ROOFING PANELS

Follow the recommendations given by the metal roofing manufacturer for the requisite sheathing. The following recommendations are typical. Tern roofing is often installed over 1 x 6 or 1 x 8 solid-wood board spaced about 2 inches apart. This allows any moisture that develops on the back of the material to dry, reducing the chances of rusting from the back side. Take care the attic has adequate natural ventilation (**9-11**).

Structural metal panels can be installed over spaced wood purlins (**9-12**). Typically 1 x 4 or 2 x 4 purlins are used; the 2 x 4 purlins give a much more rigid roof, which you will notice as you walk on it. The spacing of the purlins depends on the structural properties of the structural metal panels and the live loads common for the area; these include snow and wind. The manufacturer will have data on spacing, but distances of 16 to 36 inches are common. This type of roof is not energy efficient because the metal heats up and radiates to the building below. For barns and storage buildings this may not be a problem. For a residence, however, special consideration is needed to insulate the ceiling below and provide considerable ventilation.

Copper, steel, lead, tern metal, and zinc roofing panels are installed over solid sheathing. These panels have little strength to support themselves

COPPER FINISH   STONE COATED

WOOD GRAIN   SMOOTH FINISH

9-10 Metal roof panels simulate shingles and are available in a number of profiles and finishes. **Courtesy Metalworks Roofing Systems**

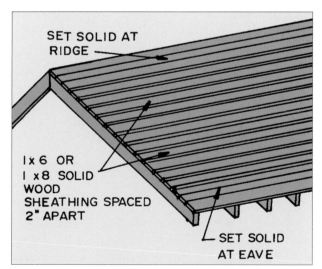

9-11 Solid-wood sheathing used with tern roofing is spaced 2 inches apart to allow air circulation.

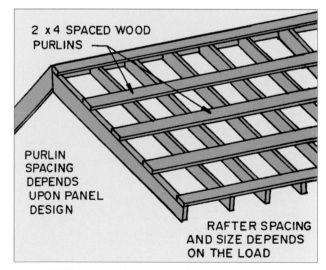

9-12 Widely spaced wood purlins can be used to support structural metal roof panels.

9-13 Structural standing-seam panels can be installed by screwing them to metal purlins that are spaced over and screwed to metal rafters. **Courtesy Kundson Manufacturing Inc.**

and will bend under loads. Plywood and OSB are often used for sheathing in residential construction.

For commercial and residential steel-framed buildings where structural standing- seam metal roofing panels are used, metal C- or Z-bar steel purlins are spaced over and screwed to the metal rafters to provide the supporting framework (**9-13**). Architectural panels are installed over metal decking as shown in **9-14**.

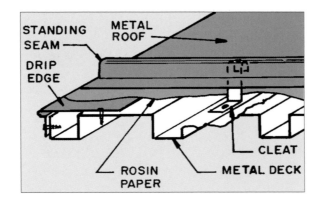

9-14 Architectural metal roofing panels are installed over metal roof decking in steel-frame buildings.

CUTTING METAL PANELS

Panels can be cut on the site by using an electric sheet-metal nibbler to cut across the panel and electric metal shears to make long cuts the length of the panel. Some panels can be cut with hand sheet-metal shears.

FASTENING
METAL ROOFING PANELS

Metal roofing panels are secured to the sheathing with special **screws** that are rust resistant (**9-15**). The head is covered with a stainless-steel cap fixed over the integral cupped hex washer head. The screw has an EPDM plastic sealing washer that bonds to the roofing material, providing a leak-proof seal (**9-16**). The screws are self-drilling, meaning the tip drills a hole through the metal and guides the screw into the wood. Nails are never used in residential construction. Nails may be used on an outbuilding where watertightness is not a major factor.

When panels are secured with screws, the screws should be placed in the trough. This pulls the roofing tight to the sheathing. If it is placed through the crown, the panel will flex due to expansion and contraction and can loosen the screw in the sheathing. When screwed directly to the sheathing, this movement is eliminated (**9-17**). Some manufacturers recommend that panels with rounded troughs (corrugated roofing) be screwed through the crown (**9-18**); the screws used have a sealing washer below the head to stop water penetration.

Most panels used in residential construction use some type of cleat, snap, or cap to secure the metal roofing. These hold the edges of the panels together, conceal the fasteners, and waterproof the joint.

9-15 It is important to secure metal roofing with drill screws that will not rust. **Courtesy SFS intec, Inc.**

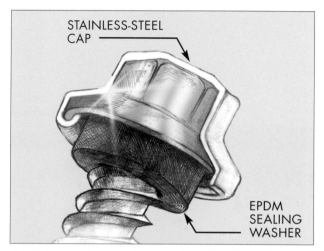

9-16 Screws for securing metal roofing have the head covered with a stainless-steel cap. It also has an EPDM sealing washer to waterproof the connection. **Courtesy SFS intec, Inc.**

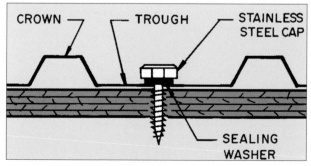

9-17 Screws are placed in the trough when securing rib-type metal roofing.

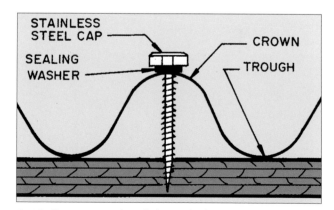

9-18 Screws are placed through the crown when installing corrugated-metal roofing.

A simple standing-seam joint has one panel edge bent over a lip on the other and the seam is secured with a rivet, as shown in **9-19**. A more secure joint uses a double-locked joint (**9-20**). Both of these seams are formed, after the panels are on the roof, by a seam-forming machine that moves down the seam and rolls the edges together (**9-21**). Notice these use cleats to secure the roofing to the sheathing and that they are covered as the seam is formed; no fasteners are exposed. Manufacturers have other seam designs they may recommend.

Another type of standing seam secures the adjoining panels to the sheathing with a cleat and then puts a seam cover over it. This is a tight-fitting cover that snaps over the panel edges. It is often referred to as a snap-on standing seam (**9-22**).

Another way to join metal roofing panels is with a batten seam. One type uses a metal channel strip between the panels. The panels are locked into this channel and the batten cap is snapped over the entire assembly (**9-23**).

Manufacturers have other forms of seam assembly available; consult them before you make an installation. Note that some manufacturers have patented connectors.

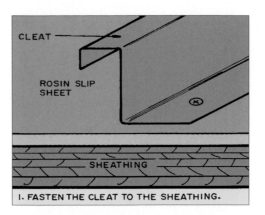

1. FASTEN THE CLEAT TO THE SHEATHING.

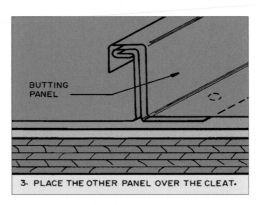

2. BUTT A PANEL TO THE CLEAT. FOLD THE CLEAT AROUND IT.

3. PLACE THE OTHER PANEL OVER THE CLEAT.

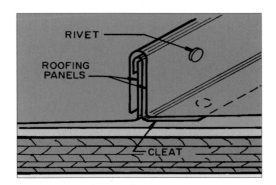

9-19 A simple standing seam.

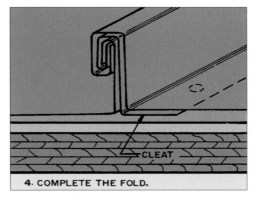

4. COMPLETE THE FOLD.

9-20 (Left, top to bottom) A double-locked standing seam.

9-21 This machine moves down the standing seam and folds the joining pieces, forming a watertight joint.

Courtesy ATAS International, Inc.

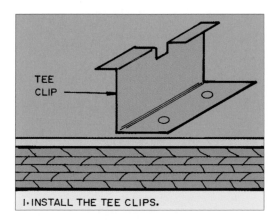

9-22 (Middle and bottom) This snap-cap standing seam uses T-clips to join the metal roof panels to the sheathing.

1. INSTALL THE TEE CLIPS.

2. BUTT ROOF PANELS TO THE TEE CLIP, BEND THE TABS OVER THEM AND INSTALL THE SNAP CAP.

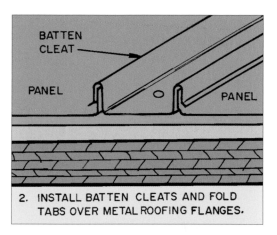

1. INSTALL THE METAL ROOFING PANELS SPACED TO ACCEPT THE BATTON CLEATS.

2. INSTALL BATTEN CLEATS AND FOLD TABS OVER METAL ROOFING FLANGES.

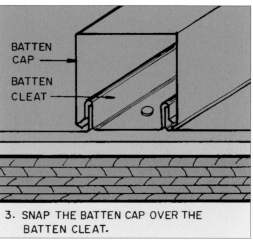

3. SNAP THE BATTEN CAP OVER THE BATTEN CLEAT.

9-23 (Above top, middle, and bottom) Ribbed metal roofing panels can be joined using a snap-on batten seam. This gives prominence to the ribbed appearance of the roof assembly.

METAL ROOFING

INSTALLING THE
METAL ROOFING PANELS

Roofers work carefully when installing metal roofing. First, it is slippery, so falls are a major danger (**9-24**). Then scratches in the exposed surface must be avoided. If a scratch occurs, it must be painted with a manufacturer-supplied touch-up paint. Roofers wear soft-soled shoes to reduce scratching and slipping. Any tools on the roof must be buffered with large foam-rubber pads made for this purpose (**9-25**).

Usually the metal rake and eave flashings are installed first. Then the ice-protection membrane and the underlayment are installed along the eave. Some roofing contractors also install it along the rake. The underlayment is red rosin paper and serves as a slip sheet to protect the ice-protection membrane. The edge flashings are secured to the sheathing as recommended by the manufacturer.

The **rake** is sealed with a metal rake cover made of the same material and color as the roofing. One design provides a standing seam along the rake, guiding water flow to the eave (**9-26**). The other provides a smooth flush edge, allowing water to flow off the rake (**9-27**). A typical rake flashing for roofs using ribbed metal roofing is shown in **9-28**.

9-24 Working on this tern metal roof is a dangerous job. Every safety precaution should be taken and cooperation between the roofers is important.
Courtesy Murton Roofing of South Carolina, Inc. and Don Wheeler, Photographer.

ROOFING MATERIALS & INSTALLATION

9-25 Take precautions to prevent scratches, dents, and other damage to the metal roofing.

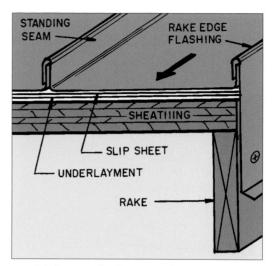

9-26 The metal rake cover directs the flow of water toward the eave.

STANDING SEAM

RAKE EDGE FLASHING

SHEATHING

SLIP SHEET

UNDERLAYMENT

RAKE

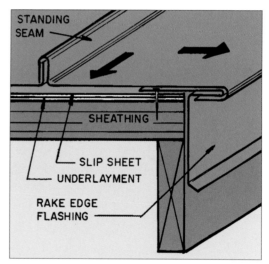

9-27 A smooth, flush rake cover lets water flow over the edge of the rake.

STANDING SEAM

SHEATHING

SLIP SHEET

UNDERLAYMENT

RAKE EDGE FLASHING

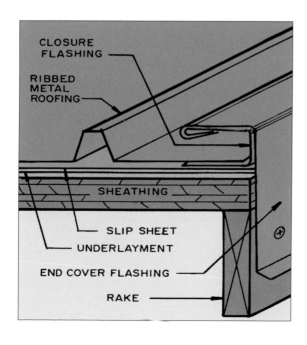

CLOSURE FLASHING

RIBBED METAL ROOFING

SHEATHING

SLIP SHEET

UNDERLAYMENT

END COVER FLASHING

RAKE

9-28 (Left) A typical rake cover for ribbed metal roofing.

Typical construction at the **eave** is shown in **9-29**. The eave flashing extends at least 3 inches up on the roof. The ice-dam membrane is required in areas where snow will occur. It should extend up the roof until it is at least 2 feet inside the exterior wall. The red rosin slip sheet is laid over it. Notice that, when the metal roofing butts the eave flashing, the end is turned under the lip on the flashing. This prevents the wind from lifting the roofing.

As the first row is installed, it is vital that it be parallel to the rake board. Run a chalk line and measure carefully before securing it to the sheathing. As additional panels are laid, be careful they are not bent somewhat, which could influence the width. Since panels may be very long, it is easy to get a bow in the overall length.

Be very careful to bend the end of the roofing over the lip on the eave flashing. Be certain the roof is secured to the rake flashing. If these are not properly done, the wind will lift the metal roofing.

The location and number of fasteners, screws, or cleats varies with the design of the metal roofing. Follow the manufacturer's specifications as to the type of fasteners and their location.

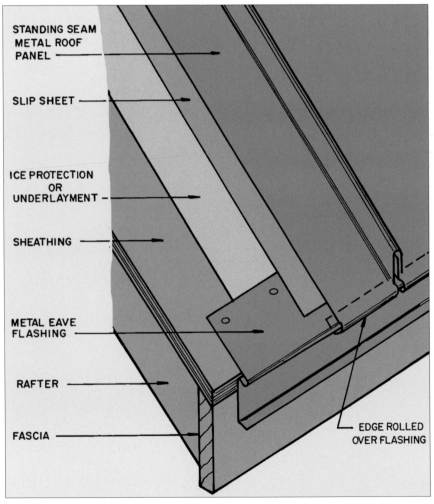

STANDING SEAM
METAL ROOF
PANEL

SLIP SHEET

ICE PROTECTION
OR
UNDERLAYMENT

SHEATHING

METAL EAVE
FLASHING

RAFTER

FASCIA

EDGE ROLLED
OVER FLASHING

9-29 One way to flash the eave when using standing seam metal roof panels.

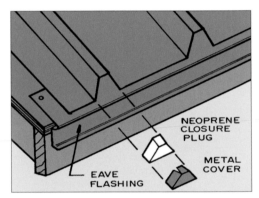

NEOPRENE
CLOSURE
PLUG

METAL
COVER

EAVE
FLASHING

9-30 Ribbed metal roofing requires that closure plugs be installed along the eave.

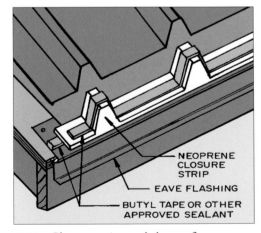

NEOPRENE
CLOSURE
STRIP

EAVE FLASHING

BUTYL TAPE OR OTHER
APPROVED SEALANT

9-31 Closure strips seal the roofing to the eave flashing and close the openings in the ribs.

If ribbed-metal roofing panels are used, the ends must be sealed along the eaves with closure plugs (**9-30**). Ordinarily a metal cover is placed over the plug. It is made of the same color metal as the roof panel. Another technique is to use a full-length closure strip and bond the panel to it with an approved sealant (**9-31**).

PANEL END LAPS

Should the length of the roof from ridge to eave be too long to allow the use of a single length of metal roofing, end laps must be made. It is important that you follow the manufacturer's directions. Some manufacturers factory-notch the vertical rib on the lower panel. It is recommended that the end laps be staggered 12 inches. Some designs line them up for aesthetic reasons. Specific details vary by manufacturer; follow their instructions.

One panel end lap design is shown in **9-32**. The upslope panel must always be on top of the panel below it. The fasteners used should be

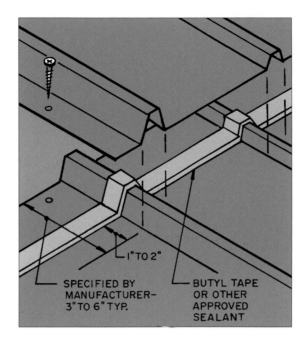

9-32 One technique used to seal end laps between metal roofing panels. Overlap the amount that is recommended by the panel manufacturer. Under some conditions a double row of sealant may be required.

behind the sealant so that they are not exposed to the weather. The lap should be at least 3 inches; however, it may be more depending on the slope of the roof. Some manufacturers supply metal roofing panels with a folded flat-seam joint already prepared. The top edge of the lower panel is held with a cleat as shown in steps one and two in **9-33**.

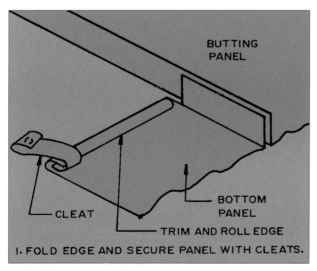

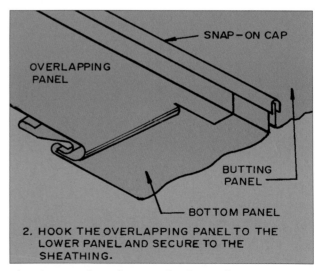

9-33 This example shows a cross seam used to connect overlapping metal-roofing panels. Cut and space as recommended by the manufacturer. **Courtesy Follansbee Steel**

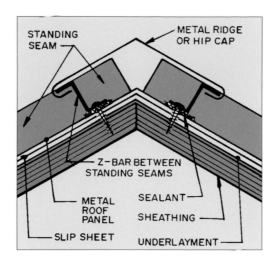

9-34 One technique used to install ridge and hip caps. The Z-bar fastener is cut and fit in between the standing seams.

STANDING SEAM · **METAL RIDGE OR HIP CAP** · **Z-BAR BETWEEN STANDING SEAMS** · **METAL ROOF PANEL** · **SEALANT** · **SHEATHING** · **SLIP SHEET** · **UNDERLAYMENT**

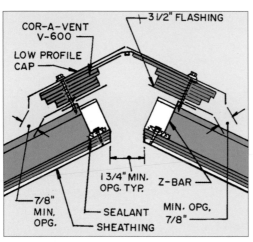

9-35 A ridge vent installed on a standing-seam metal roof. Similar designs are available for ribbed and other metal roofing panels. **Courtesy Cor-A-Vent, Inc.**

COR-A-VENT V-600 · **3 I/2" FLASHING** · **LOW PROFILE CAP** · **I 3/4" MIN. OPG. TYP.** · **Z-BAR** · **7/8" MIN. OPG.** · **SEALANT** · **SHEATHING** · **MIN. OPG. 7/8"**

INSTALLING RIDGE CAPS & HIP CAPS

Ridge caps may be venting or non-venting. The venting type gives the advantage of moving hot attic air out from below the roof and in the winter removing moisture that may have collected there. Metal roofing manufacturers have a number of ridge vents available. Each has a slightly different design. A general type similar to those in common use is shown in **9-34**. It uses a metal Z-bar between standing seams. The vent is screwed to the sheathing. The preformed metal ridge cap is secured to it by bending the edges around the bar flange.

A vented ridge cap installed on a standing-seam metal roof is shown in **9-35**; it secures the vent cap to a Z-bar that is secured to the sheathing. The vent space contains a special honeycomb section through which the air passes. Various manufacturers have similar ridge cap systems.

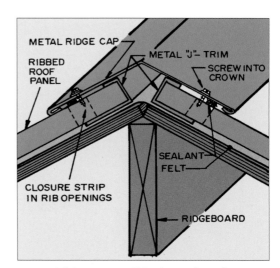

9-37 A hip cap on ribbed metal roofing can be secured at the ridge with J-trim and sealed with a neoprene or rubber closure strip.

METAL RIDGE CAP · **RIBBED ROOF PANEL** · **METAL "J" TRIM** · **SCREW INTO CROWN** · **SEALANT FELT** · **CLOSURE STRIP IN RIB OPENINGS** · **RIDGEBOARD**

9-36 A continuous hip cap used on metal-panel roofing.

The hip is covered with a continuous hip cap (**9-36**). A typical installation detail for ribbed roofing is shown in **9-37**. The manufacturer supplies a metal J-trim and rubber or plastic closure plugs. The preformed hip cap is screwed to the metal crown of the roofing.

FLASHING

Flashing a metal roof installation takes some techniques that are different than those methods commonly used with other types of roofing materials. Always observe the roofing manufacturer's instructions and use the flashing materials as specified or provided. Flashing includes valleys, butting walls or chimneys, and around pipes.

FLASHING VALLEYS

Valleys are flashed with the same metal used for the roof panels (**9-38**). Flashing techniques are about the same as discussed for other roofing materials. The valley has the edges crimped, into which clips are placed and screwed to the sheathing (**9-39**). The valley should be 18 inches wide. End laps are 4 inches and sealed with a double row of butyl tape or other sealant.

9-38 This valley is on a roof using standing-seam metal roofing. It is made of the same material and color as the roofing.

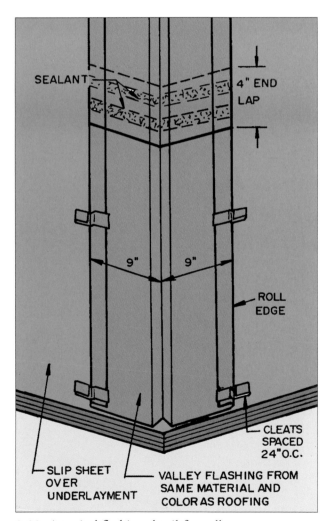

9-39 A typical flashing detail for valleys on metal-finished roofs.

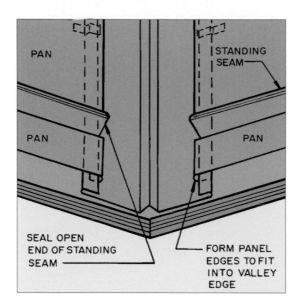

9-40 The edges of the pan are formed to secure the standing-seam metal roofing to the valley.

A detail for installing the roofing at the valley when the roof uses standing seams is in **9-40**. The edges of the roof panels are cut on the angle of the valley and have the end formed to fit over the rolled edge of the valley.

Ribbed roofing will have large openings where the raised ribs touch the valley. These must be sealed with closure plugs cut from a material, such as neoprene. The trough is sealed to the flashing with butyl tape or some other approved sealant (**9-41**).

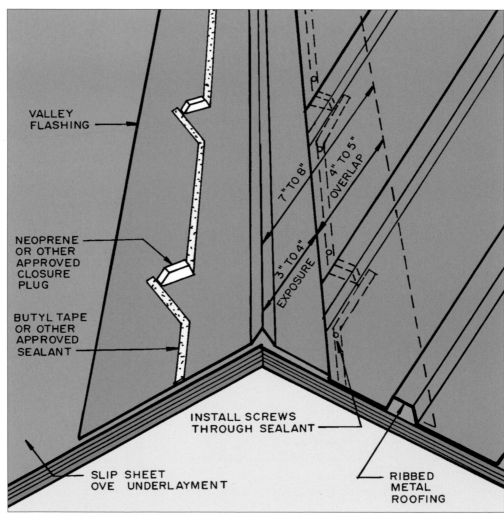

9-41 The edge of the ribbed metal roofing along the valley requires closure plugs in the rib openings and that the trough be sealed to the valley.

FLASHING BUTTING WALLS OR CHIMNEYS

When the metal roofing meets a vertical wall, the design of the flashing is somewhat different, depending on whether it runs perpendicular to or parallel with the wall.

Typical details for a perpendicular situation are shown in **9-42**. The end of the metal panel between the standing seams is bent up the same height as the seam. A Z-bar is screwed through the panel into the sheathing. The L-shaped flashing has a rolled edge fitting over the flashing.

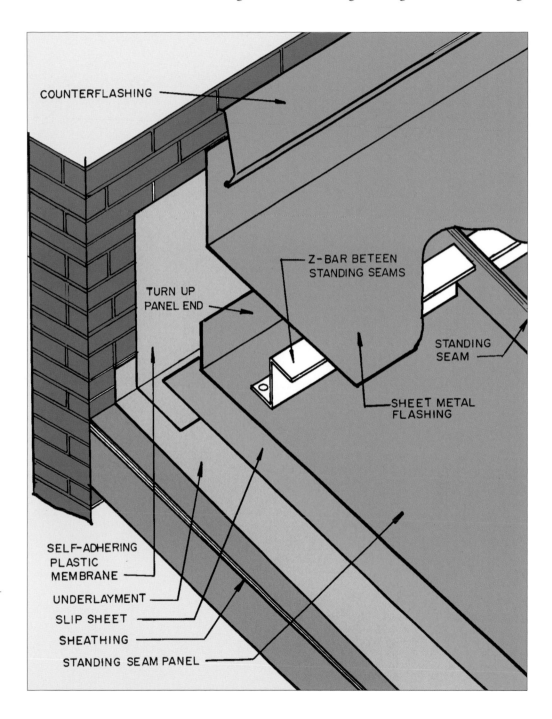

9-42 Flashing details when a metal standing-seam roof runs perpendicular to a wall.

9-43 When the metal roof meets a wall finished with some type of siding, the siding overlaps the flashing.

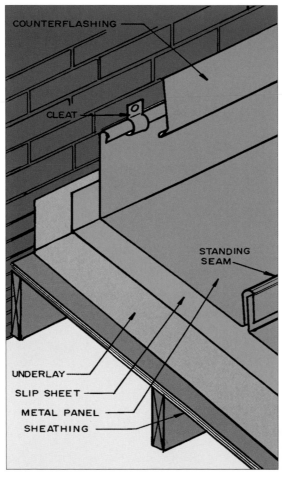

COUNTERFLASHING

CLEAT

STANDING SEAM

UNDERLAY
SLIP SHEET
METAL PANEL
SHEATHING

9-44 Flashing details when a standing-seam metal roof runs parallel with a wall.

Notice also that in **9-42**, on the previous page, the roof underlayment is lapped up the wall and covered with a self-adhering plastic membrane. The assembly is topped off with counterflashing. If the wall has some type of siding, the siding overlaps the flashing so that the flashing is concealed below as shown in **9-43**. The siding is kept 1 inch above the flashing on the roof.

The construction of the flashing along walls parallel with the run of the panel is much like the perpendicular flashing, except that the metal roofing is bent and laid up the wall to form a trough for handling the flow of water (**9-44**). If the wall has some form of siding, the siding is laid over the flashing as shown in **9-45**. The siding is cut so it is 1 inch above the surface of the flashing on the roof.

9-45 The edge of the metal roofing is laid up on the wall and the siding is installed over it.

INSTALLING PIPE FLASHING

The intersection of a pipe that protrudes through a roof with the roofing presents a major waterproofing problem. The pipe shown in **9-46** is on a 100-year-old house that is having a new metal roof installed over the existing old roof. Notice the deterioration of the pipe flashing, which will be replaced with some form of newer boot or flashing. Another problem to consider is the need to allow for expansion and contraction of the pipe and flashing. Some installers use a rubber boot to seal the pipe. In any case, the hole made in roof must allow room for the pipe to expand.

9-46 This pipe flashing is on the roof of a very old house. As the house is reroofed it will be replaced with a newer flashing unit.
Courtesy Murtow Roofing of South Carolina, Inc. and Don Wheeler, photographer

The rubber boot is bonded to the metal panel and secured with screws (**9-47**). If the pipe hits the roof where a boot cannot seat or if the roof is corrugated, some installers seal around the pipe with roofing cement and build it up in a cone 3 or 4 inches high. When it sets up, it forms a boot and allows for thermal expansion and contraction.

Another type of boot for flashing a pipe has a flexible aluminum compression ring that permits it to be sealed and screwed to almost any rib profile (**9-48**). For dealing with ribs or other raised surfaces, this moldable ring allows a seal to be formed between the pipe and the roof surface. There are a number of other products available for flashing pipe penetrations.

9-47 (Middle and bottom) One type of pipe flashing seals a flashing unit to the surface of the metal-roofing panel.

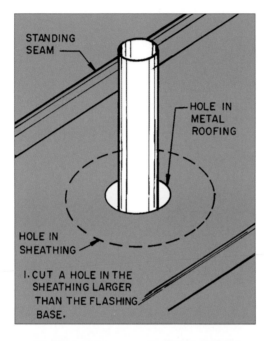

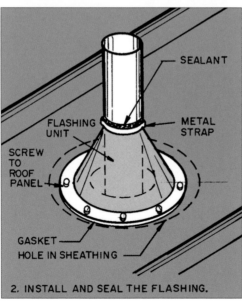

WORKING WITH METAL SHINGLES

Metal shingles simulate clay tile, slate, and wood shakes (**9-49**). The image, profile, and texture are molded into the metal. They have various coatings, as described earlier for panel products. Some manufacturers offer stone-coated steel shingles (**9-50**).

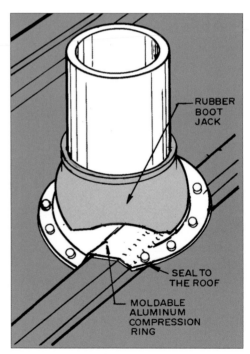

9-48 This pipe flashing has a flexible rubber boot that is sealed to the roof by a moldable-aluminum compression ring. The ring can be formed to fit over ribs and other raised surfaces.

Shingles are available as individual shingles, such as the unusual diamond-shaped metal tile shown in **9-51**. Another approach is the standing-seam shingle shown in **9-52**. It provides a staggered-end-seam design, giving an attractive appearance to the roof. These are single shingle products.

9-51 These 15¾-inch diamond-shaped metal roof tiles have an expanded polystyrene backer board that gives strength to allow for foot traffic during installation. They have edges turned to seal against rain penetration. **Courtesy ATAS, International, Inc.**

9-49 This roof has been covered with a stone-coated steel roof that provides the appearance of wood shakes. **Courtesy Tasman Roofing, Inc.**

9-50 This stone-coated steel roofing simulates asphalt shingles. It is much lighter and has a 50-year limited warranty. **Courtesy Tasman Roofing, Inc.**

9-52 This roof is being finished with a standing-seam shingle. The shingles are 36 inches long and are installed with the end joints staggered to provide a decorative feature. **Courtesy ATAS International, Inc.**

Another type of metal shingle has the images stamped into a panel 3 or 4 feet long and 12 to 20 inches wide. Each panel has one, two, or three courses of the shingle image repeated along the length of the panel (**9-53**). The roof shown in **9-54** has been finished with stone-coated steel shingle panels. Tile roofing panels duplicate the profile and color of clay tile (**9-55**). They are lightweight and have a long warranty.

Another shingle product is available in panels from 2 to 20 feet long and contain six corrugations in the 36-inch width (**9-56**). The finished installation with matching fascia and rake trim provides an attractive, lightweight roof with a long life (**9-57**).

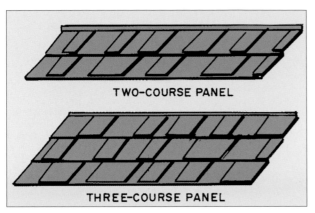

9-53 Metal shingles are available in panels containing one, two, or three courses stamped in a repeated series of profiles along the length.

9-54 This roof has been finished with aluminum-zinc-coated steel with an acrylic-bonded, stone-chip finish. It is highly resistant to nail penetration and high winds. **Courtesy Tasman Roofing, Inc.**

9-55 Metal roofing molded and finished to represent clay tile provides a durable, lightweight alternative to the actual clay-tile roofing. **Courtesy Tasman Roofing, Inc.**

9-56 This metal roofing panel is formed from 26-gauge steel and is available with several types of finish. It is installed over plywood sheathing with a 30-pound felt underlayment.
Courtesy Met-Tile, Inc.

INSTALLING METAL SHINGLES

Metal roofing shingles are installed using techniques similar to those shown earlier in this chapter for metal roofing panels. Each manufacturer has detailed installation manuals for its products and some also have video tapes available. Consult these types of sources before attempting the installation.

9-57 The finished installation provides an attractive finished roof that has a long life. **Courtesy Met-Tile, Inc.**

Built-Up, Modified-Bitumen & Roll Roofing

Built-up roofing (BUR) is one of several types used on low-slope roofs and is widely used on commercial buildings. In residential work BUR is used on low-slope roofs in warm climates with moderate rainfall. Basically it consists of building layers of reinforcing felt coated with a bitumen. While there are several finish coats, the one used in residential housing has a top coat of gravel embedded in the last coat of bitumen. This protects the assembly and gives the roof an attractive color and texture (**10-1**).

A variation is constructed with fabric-reinforced sheets formulated with modified-bitumen that has improved properties relative to standard bitumen. Modified-bitumen roofing is used on low-slope roofs. An alternative to both is roll roofing, which is used not only on flat and low-slope roofs, but also on steep-slope roofs. Roll roofing is an easy-to-install, low-cost roofing product, but it does not have a pleasing appearance and tends to leak more easily than other roofing materials.

CONSIDERING BUILT-UP ROOFING

Installation of BUR is a hot, hard job. Heating the bitumen, pumping it to the roof, and spreading it is hot, difficult work. While it has been used for many years and produces a quality roof covering, it is rapidly being replaced by singly-ply membrane roofing materials. These are covered in Chapter 11.

The design of the assembly and the selection of materials used is the work of someone experienced in BUR construction; the homeowner should not attempt to do this job. In this chapter each material will be identified by a generic name. For example, when a term like bitumen is used, there are several formulations available from which to choose. The discussion will relate to residences that have solid-wood or plywood

sheathing with a slope of 3:12 or less and the top finish coat to be an aggregate or mineral. Be aware that built-up roofs on commercial buildings are typically installed over steel and concrete decks that require additional explanation not covered here.

SHEATHING

Solid-wood, plywood, and OSB (oriented-strand board) provide a satisfactory base upon which to lay the built-up plies. It is important that sheathing be thick enough to carry the load between rafters. It should have all edges supported and any holes or damage repaired to get a flush panel. Typically, plywood sheathing will be ½-inch thick and solid-wood boards 1-inch thick nominal (¾-inch dressed thickness). Rafter spacing will determine these thicknesses.

HOT BITUMENS

The most frequently used bitumens are **asphalt** and **coal-tar pitch**. Coal-tar pitch is more expensive; it is recommended for use on roofs that have slopes below 1:12. There are several types of coal-tar designed for different applications.

10-1 Built-up roofs have a layer of rock or mineral aggregates laid in the top coat of hot bitumen.

Asphalt is available in four types. **Type I** asphalt is used on perfectly level roofs up to slopes of ½:12. **Type II** is used on roofs with slopes between ½:12 and 1½:12. **Type III** is less likely to flow when the roof gets hot in the summer and is used on roofs with slopes between 1:12 and 3:12. **Type IV** is used on steeper roofs and in areas where high temperatures exist all year; it is used on roofs with a slope of 2:12 to 6:12. The danger in periods of high temperatures is that the asphalt may soften and run, allowing the felts to slide down the slope.

COLD-APPLIED BITUMENS

There are several cold-applied bituminous materials. A bituminous grout made of fine sand and a bitumen can be heated and poured or applied with a trowel when cold. Various asphalt emulsions are thin enough to be sprayed. Cutback bitumens are in liquid form. The bitumen is thinned with light oils and solvents. It becomes solid when the solvent or water in them evaporates. Asphalt cements are very thick, pure asphalts that are applied with a trowel; they are used to seal flashing and laps in other materials.

ROOFING FELTS

Organic felts saturated with asphalt or cold-tar bitumen are available. The felt selected should be compatible with the hot melt coatings that will go between felts.

Saturated and coated felts are available with organic or inorganic base materials. They have a light coating of a very fine mineral sand which prevents the material from sticking together when it is in a roll.

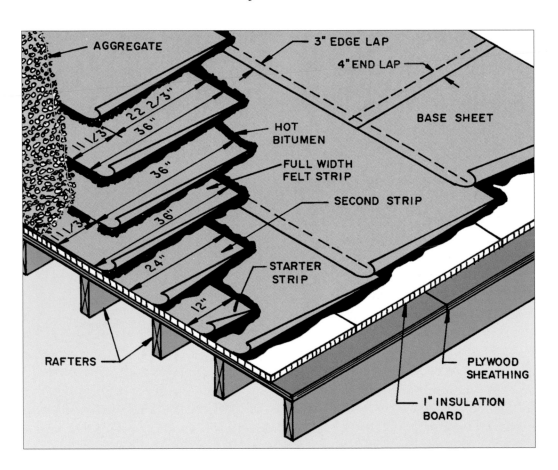

10-2 Typical construction of a built-up roof.

Impregnated felts are lightweight materials. The surface is not completely coated with asphalt. They are porous and allow vapors developed as the membrane is applied to be vented.

SURFACE AGGREGATE

Aggregate is applied in the last top coat of hot bitumen. It protects the roof when walked on and reduces the effects of weathering. It serves as a ballast to help the roof resist uplift by the wind. When light color aggregates are used, the heat from the sun is reflected back off the roof, reducing its temperature. The selection of aggregate gives color to the roof.

Aggregates commonly used include gravel, crushed stone, and crushed slag. Lightweight aggregates should be avoided because they are more likely to be washed off the roof.

INSULATION BOARD PRODUCTS

The roof membrane can be laid over rigid insulation boards that are secured to the roof sheathing. They reduce heat flow through the roof. A single or double layer of insulation board can be used.

There are several types of insulation board available. The person designing the entire built-up membrane will make the selection as to which materials should be used.

Expanded perlite insulation board is made of expanded volcanic rock and synthetic binders. **Polyisocyanurate (ISO) insulation board** is formed from a plastic foam produced by the reaction of a polyol and polymeric polyphenyl-polymetholyne diisocyanurate (PMOI). It has gray facer sheets on each side. It should be covered with a layer of fiberboard or perlite board insulation before the built-up membrane is laid.

Fiberboard panel insulation is made of cellulose, wood, and vegetable fibers bonded to form a rigid panel. It is used as the top insulating layer over various foam-plastic boards. **Composite-board insulation** has a plastic-foam core with layers of fiberboard or perlite bonded to it; it is good for hot-mopped bitumen applications.

THE BASE SHEET

The base sheet is installed directly on the sheathing. It serves as the surface on which the rest of the layers of bitumen and felt are laid and serves as a vapor barrier. This protects insulation, if used, and the built-up membranes from damage by moisture developed inside the residence. It can be a coated sheet of builder's felt. This sheet is coated with bitumen on both sides and has a coating of nonsticking fine mineral granules. A vented base sheet consists of a fiberglass mat impregnated with asphalt. It has small holes spaced uniformly several feet apart across the sheet. They let moisture escape as the installation is being made. Some designers require a layer of fiberboard or perlite be placed over vented base sheets to reduce the chance that trapped moisture and air may cause blisters in the membrane. There are other similar products available.

THE BUILT-UP ROOFING MEMBRANE

While there are many ways to build the membrane, the following is typical for residential construction. The membrane will typically include builder's felts as a base sheet over the plywood sheathing, on which alternate layers of hot bitumen and felt are laid, finished off with a top coating of hot bitumen and an aggregate. A typical BUR construction is shown in **10-2**. If the sheathing is wood boards, nail a layer of rosin-sized sheathing paper over them to keep the hot bitumen from dripping between them onto the construction below. The BUR construction in **10-2** can be used when plywood or OSB sheathing is used.

The base sheet is nailed as shown in **10-3**. The edge lap is 3 inches and is nailed at least every 8 or 9 inches. The nails in the field of the base sheet are in rows 12 inches apart and nails are placed at least every 18 inches. The sheets are end-lapped at least 4 inches. Use a nail with a large head long enough to penetrate the sheathing.

Next the plies of felt are laid in hot bitumen. Various numbers of plies of felt are used with three, four, and five plies most common. A typical installation will use 15-pound asphalt-saturated, perforated organic felt laid in a bed of hot asphalt.

Notice in **10-2**, on page 178, the first strip at the eave is 12 inches wide. Then a 24-inch sheet is laid in bitumen over it. All following sheets are the full 36-inch width of felt with hot bitumen mopped between coats. The felts are rolled out and smoothed with a large broom. The bitumen

coverage for each layer should be complete, leaving no uncoated felt exposed.

Sometimes sheets of rigid insulation are applied over the sheathing before the membrane is built (**10-4**). These preformed-insulation-board products reduce heat flow. The built-up membrane is laid on top of them. They are laid with the end joints staggered. The edge joints should be in a continuous straight line. The long dimension of the insulation panel should run perpendicular to the roof slope, as shown in **10-5**. The insulation is secured to the wood or plywood sheathing with nails specified for this purpose. If a second layer of insulation is required, it is laid in hot asphalt or an approved adhesive. Panels that are 2 x 4 feet and 3 x 4 feet are typically nailed in each corner and in the center of the long edge. Larger panels, such as 4 x 8 feet, will typically have a grid of nails with three on the 4-foot side and four on the 8-foot

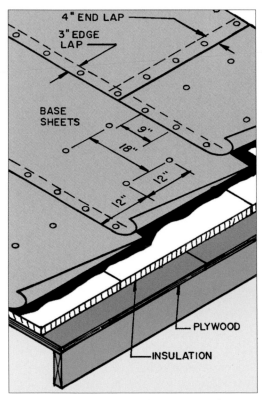

10-3 Recommended nailing pattern for the base sheets.

10-4 Rigid insulation panels are laid in a bed of hot bitumen that has been applied to the roof sheathing. **Courtesy Johns Manville Roofing Systems**

ROOFING MATERIALS & INSTALLATION

side and nails aligned with these on the interior of the panel on sheets along the perimeter of the roof (**10-6**).

After the felt layers or insulation are in place the top is coated with hot bitumen (**10-7**) and the aggregate is spread.

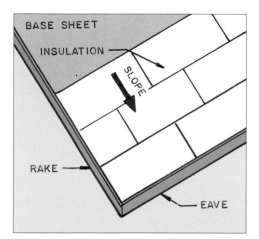

10-5 The insulation panels are installed with the long dimension perpendicular to the slope of the roof.

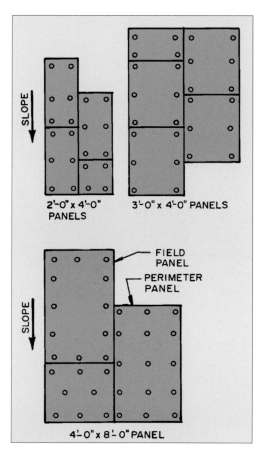

10-6 Typical nailing patterns for securing rigid insulation panels to nailable sheathing.

10-7 The top coat of hot bitumen is applied over the insulation or builder's felt. Then the aggregate is apread over it. **Courtesy Johns Manville Roofing Systems**

10-8 The top ply is hot-mopped with bitumen, over which the membrane is laid. **Courtesy Johns Manville Roofing Systems**

MODIFIED-BITUMEN MEMBRANES

Modified-bitumen membranes are fabric-reinforced sheets formulated with polymer modified asphalt and other agents to improve the properties of the bitumen. Modified-bitumen roofs provides a neat, durable finish, such as seen in the opening photo on page 176. The asphalt modifier may be polypropylene (APP) or styrene butadiene styrene (SBS). The membranes are laid in layers of two or more sheets. On residential construction they are laid over a plywood or OSB sheathing that is covered with a base sheet, serving as a vapor barrier to retard the flow of moisture from below the roof.

10-9 Roll the membrane over the hot bitumen. Be careful that air bubbles do not form under the membrane. **Courtesy Johns Manville Roofing Systems**

The first sheet is bonded to the base sheet with hot bitumen or by heating the sheet with a propane torch, which softens it enough so that it bonds to the base sheet. Using the torch method on a wood deck is questionable because of the danger of setting it on fire. Some prefer to use cold-process adhesives; they are safe and easier to apply. The torch method is widely used on concrete and steel roof decks.

The top surface can be a layer of hot bitumen with a stone or other mineral aggregate spread over it, a modified-bitumen cap sheet which could have a factory-applied granule surface coating, a smooth finished surface or metal foil bonded to it (**10-8** and **10-9**).

A total system is the same as described for a built-up roof. It consists of the roof sheathing, a base sheet vapor barrier, rigid insulation (if desired) multiple plies of modified-bitumen sheets, and some type of surface finish coating.

FLASHING

Roofing contractors use various methods to flash the eave on residential roofs to be covered with built-up roofing. The key is to seal the edge so that no water leaks in onto the sheathing.

The felt base sheet is laid over the edge of the fascia. Then the built-up plies are laid to the edge (**10-10**). The metal flashing is laid over the roofing, sealed to it, and nailed to the roof sheathing. Then two or more layers of felt or other type of membrane are laid over the flashing and out onto the roof membrane. They are laid in hot

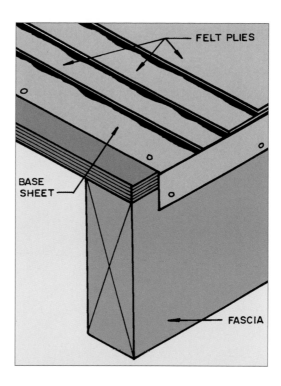

10-10 To flash the eave lap the base sheet over the edge and lay in the plies of felt in hot bitumen.

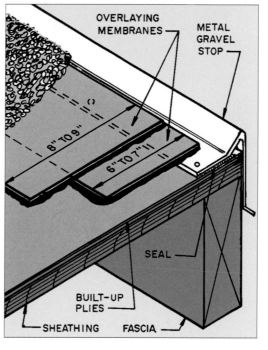

10-11 After the plies are laid, install the gravel stop and seal it with two layers of an approved membrane. Flood with hot bitumen and spread the gravel.

bitumen (**10-11**). The top membrane is flooded with hot bitumen and gravel is spread over the entire roof. Edge flashing on commercial buildings will differ from this example.

10-12 One way a built-up roof can be flashed when it meets a masonry wall.

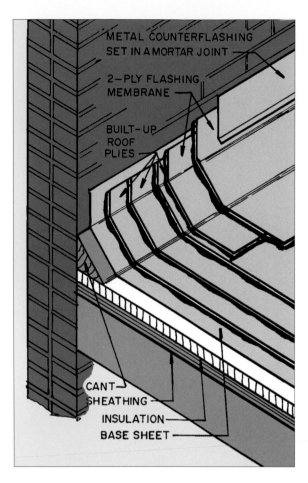

When the roofing meets a masonry vertical wall, a construction like that shown in **10-12** is typical. The cant strip and masonry wall are mopped with hot bitumen and the membrane layers are lapped up and bonded one by one over each other up the remaining length of the wall. The metal counterflashing is set in a mortar joint and can be sealed to the membrane. There are a number of different designs for applications on commercial buildings.

Pipe penetrations are sealed with commercially available prefabricated cover. It is placed over the pipe and sealed to the roof membrane (**10-13**). The flange is then primed and covered with several plies extending out around the pipe. The

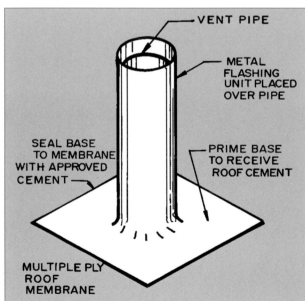

10-13 Lay the pipe flashing over the pipe and cement to the roof membrane.

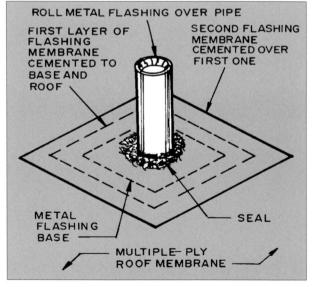

10-14 Prime the flashing base and lay two felt membranes over it, bonding them to the roof membrane and each other with hot bitumen.

junction between these plies and the pipe is sealed with a compatible roof cement (**10-14**). The roof is flooded with hot asphalt and the gravel top layer is spread.

ROLL ROOFING

Roll roofing is used on flat, low, and steep slope roofs. It is a product that is easy to install at a low cost; however it has a short life, does not have a pleasing appearance, and tends to leak more easily than other roofing materials. It is most often used on storage buildings and sheds; however, it can be effective on small porches where the roof is not visible.

Roll roofing is made with either organic felt or fiberglass mats as the base material. A bituminous coating is applied to this base, forming the exposed surface. It is available in three types: smooth-surfaced, mineral-surfaced, and mineral-surfaced selvaged-edged (**10-15**). Smooth-surfaced roll roofing has both sides covered with a fine talc or mica to keep the surfaces from sticking when it is made into rolls. Mineral-surfaced roll roofing has mineral granules in a range of colors rolled into the surface. This is more attractive than the smooth-surfaced product. The granules also help by protecting the bitumen from damage from the UV rays of the sun.

Mineral-surfaced selvage-edge roll roofing has the same construction as the mineral-surfaced rolled product but only 17 inches of the 36-inch width of the roll is covered with mineral granules. This produces a 19-inch-wide selvage edge which is used for lapping the next layer forming a two-ply covering.

10-15 Commonly available types of roll roofing.

INSTALLING ROLL ROOFING

Smooth-surfaced and mineral-surfaced roll roofing may be installed parallel with the eave or perpendicular to it, as shown in **10-16** and **10-17**. These installations provide single coverage. Remember, as with other types of roofing materials, each course must be kept straight. As you lay out the various courses run a chalk line to check each one.

Use galvanized roofing nails long enough to go at least ¾ inch into the sheathing. In the summer the rolls are warm and unroll easily. In cold weather, below 45°F, keep the rolls in a warm room and remove and use them quickly to prevent cracking the surface.

Always use the lap or asphalt cement recommended by the manufacturer. It must also be stored in a warm room.

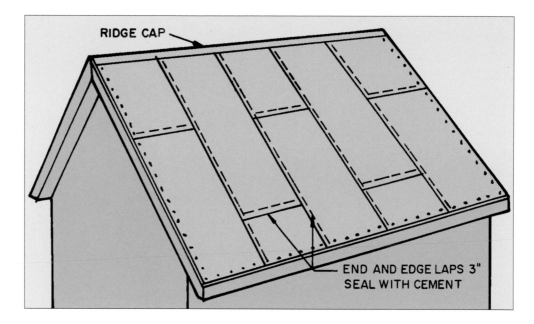

10-16 Roll roofing can be installed perpendicular to the eave.

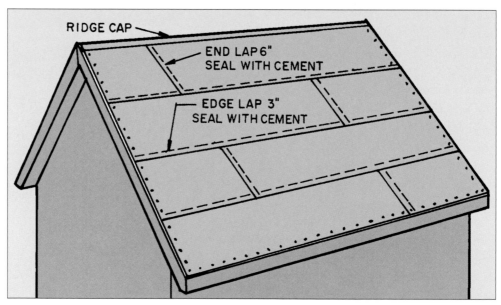

10-17 Roll roofing can be installed parallel with the eave.

EXPOSED-NAIL
INSTALLATION METHOD

The exposed-nail method is the easiest and fastest installation method; however, it is not as durable or watertight as other methods.

Begin by laying the first sheet along the eave and rake. Allow it to overlap about ⅜ inch. Use a chalk line to get it straight. Nail all edges as shown in **10-18**. The top edge nails are spaced about 18 inches apart. When the overlapping sheet is nailed, this top edge will be secured by the nails in the overlapping piece. Nail all the edges. The next sheet will overlap 3 inches along the edge. Any end laps are sealed with cement and nailed. Seal the edge laps and nail with staggered nails 3 inches apart.

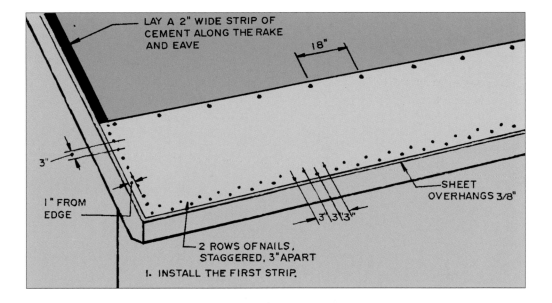

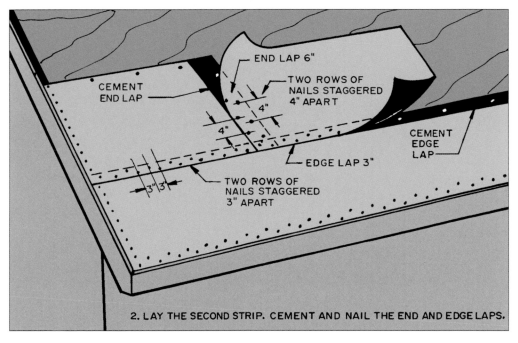

10-18 This roll roofing has been installed for single-thickness coverage parallel with the eave using the exposed-nail method.

CONCEALED-NAIL
INSTALLATION METHOD

This method provides a better-looking roof with a longer life than the exposed-nail method. Install metal drip edges on the rake and eave. Then install 9-inch wide strips along the rake and eave, allowing them to overhang ⅜ inch (**10-19**).

Place the first full-width strip along the eave and rake, allowing it the same ⅜-inch overhang. Nail the top edge to the sheathing, with two rows of staggered nails 4 inches apart. Lift the sheet along the rake and eave and coat the edge strip with cement. Press the sheet into the cement, embedding it fully.

10-19 This roll roofing has been installed using the concealed-nail method. It provides a better looking and more watertight roof than the exposed-nail method.

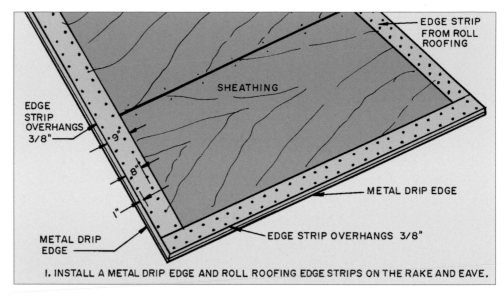

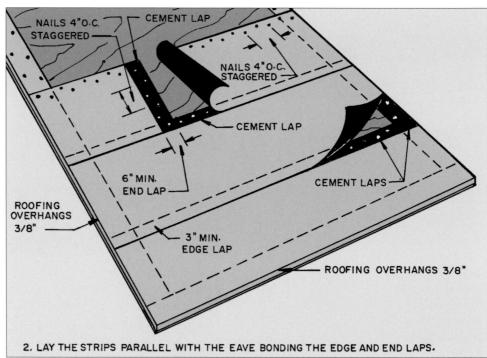

ROOFING MATERIALS & INSTALLATION

Snap a chalk line at least 3 inches below the top edge. This locates the edge of the overlap. Install the second strip in the same manner. The overlapping sheet is pressed into the cement on the overlap.

If the sheet does not reach the full width of the roof, prepare an end lap (**10-19**). It should be at least 6 inches. The bottom sheet is nailed to the sheathing with two rows of staggered nails spaced 4 inches apart. Coat the overlap area with cement and press the top piece into it.

INSTALLING MINERAL-SURFACED SELVAGE-EDGE ROLL ROOFING

Double coverage is provided by using mineral-surfaced selvage-edge roll roofing. The roofing roll has a 17-inch-wide mineral-covered surface that is exposed as well as a 19-inch uncoated selvage edge.

Prepare the roof by installing metal drip edges on the eave and rake. Cut off the 19-inch selvage from a roll and install it along the eave (**10-20**). Allow it to overhang the eave and rake about ⅜ inch. Nail it to the deck along the top and bottom edges.

Next apply a coating of roofing cement over the entire selvage-edge. Brush or trowel it out to a thin uniform coating. Lay the first sheet over this, lining up the edges with the selvage sheet. Nail it to the sheathing with two rows of nails in the selvage as shown in **10-20**.

Lay the rest of the courses so they overlap the 19-inch selvage area. First nail the selvage area as just described. Then lay the sheet back and apply cement to the entire selvage area below. Lower the sheet over the selvage and carefully press it in place. Be careful that air pockets do not collect between the sheets as you join them together.

End laps are 6 inches wide. They are bonded with a layer of cement and nails. The lower sheet is nailed along the edge; bond the lapping sheet with roofing cement.

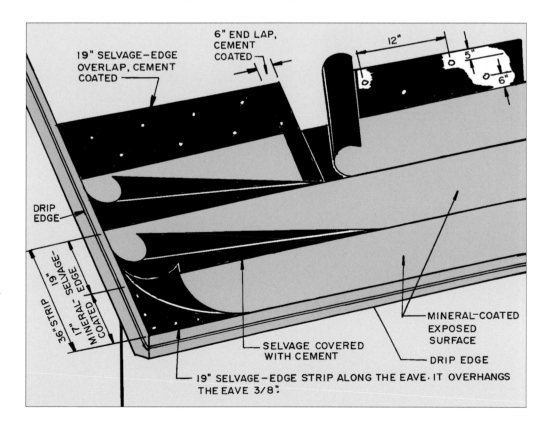

10-20 This installation uses mineral-coated, selvage-edge roll roofing to produce double coverage.

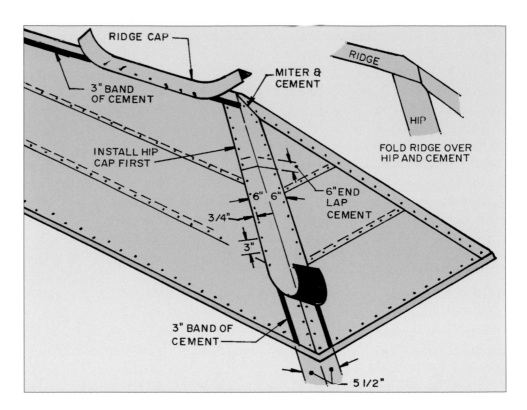

10-21 This ridge-and-hip installation leaves the nails exposed.

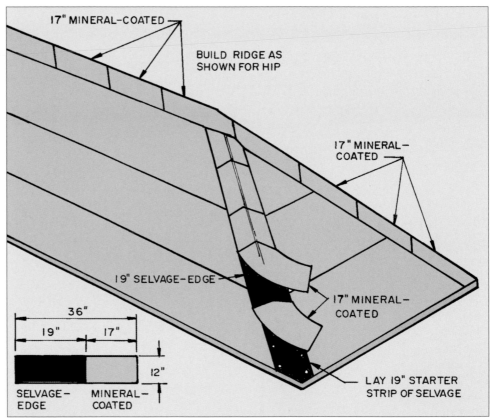

10-22 This ridge and hip installation is double covered using roll roofing that has a 19-inch selvage edge.

ROOFING MATERIALS & INSTALLATION

INSTALLING RIDGES & HIPS

The easiest, but least effective, way to install ridge-and-hip coverage is to cut strips of roll roofing 12 inches wide and the length of the ridge or hip. Run chalk lines 5½ inches on each side. Spread 3-inch bands of cement inside these lines. Lay the 12-inch strip over the ridge or hip and bend into the cement. Nail along the edges to hold it in place. This leaves the nails exposed. Install the hip first. The ridge must overlay it where it meets the hip (**10-21**).

Nails can be concealed and double coverage obtained by using mineral-coated roofing with 19-inch-wide selvage. Cut the 36-inch-width roofing into 12-inch width strips, yielding pieces 12 inches wide with a 19-inch selvage section and a 17-inch mineral-coated section.

Begin by laying a 19-inch section of selvage at the end of the ridge or hip. Cement and nail in place. Then lay pieces of the 36-inch-long strips up the hip or along the ridge by bonding to the selvage part with cement and nailing the selvage to the sheathing (**10-22**).

Overlap the hip flashing at the ridge with the ridge flashing.

INSTALLING A VALLEY

Lay an 18-inch-wide strip of roll roofing with the mineral coating face down against the sheathing. Then lay a 3-inch band of roofing cement down each side. End overlaps should be 6 inches and cemented. Now lay a full width sheet of roll roofing over this with the mineral-coated side up. Press into the cement and nail along the edges. End laps are 12 inches and are cemented and nailed. Finally, lay the roll roofing, starting at the eave and working toward the ridge (**10-23**).

Begin by running a chalk line locating the center of the valley and other chalk lines marking the width 6 inches on each side. Lay the sheets over the chalk lines and cut on the angle. Cement the edge to the valley material. Nail the sheet on the top edge no closer than 9 inches to the center of the valley.

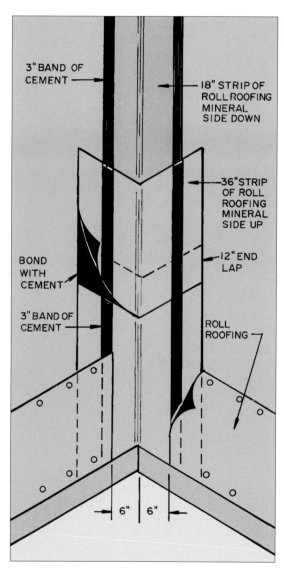

10-23 This is a typical valley flashing detail that uses two layers of roll roofing.

SEALING
VENT PIPES

For single coverage roofing first cut a hole in the roll roofing and lay it over the pipe. Cement it to the roofing and nail through it to the sheathing. Space the nails every 2 inches. Finally, cover the flashing base with roofing cement (10-24).

For double-coverage roofing cut a hole in the first sheet of roofing and slip over the pipe. Coat the area around the pipe with roofing cement. Place the prefabricated pipe vent flashing unit over the pipe. Cement and nail it through the first sheet to the sheathing. Apply roofing cement over the flashing base and out on the surrounding roofing several inches. Then cut a hole in the sheet for the second layer and lay it over the flashed pipe. Press firmly into the cement around the pipe (10-25). Seal the opening around the pipe.

10-24 Pipes can be flashed by placing a metal flashing unit over them and cementing and nailing the base to the sheathing.

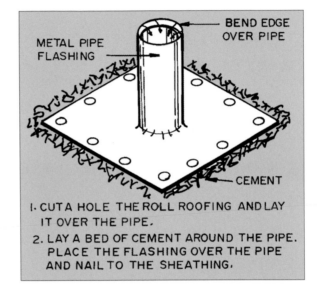

BEND EDGE OVER PIPE

METAL PIPE FLASHING

CEMENT

1. CUT A HOLE THE ROLL ROOFING AND LAY IT OVER THE PIPE.

2. LAY A BED OF CEMENT AROUND THE PIPE. PLACE THE FLASHING OVER THE PIPE AND NAIL TO THE SHEATHING.

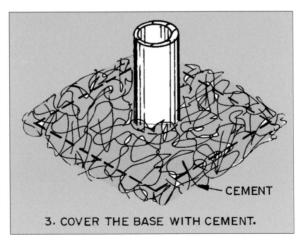

CEMENT

3. COVER THE BASE WITH CEMENT.

10-25 Steps for flashing a pipe when the roof has double coverage with roll roofing.

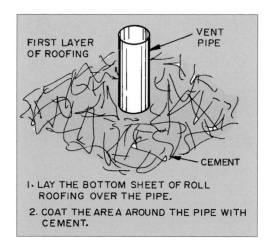

FIRST LAYER OF ROOFING
VENT PIPE
CEMENT

1. LAY THE BOTTOM SHEET OF ROLL ROOFING OVER THE PIPE.
2. COAT THE AREA AROUND THE PIPE WITH CEMENT.

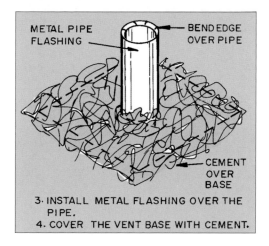

METAL PIPE FLASHING
BEND EDGE OVER PIPE
CEMENT OVER BASE

3. INSTALL METAL FLASHING OVER THE PIPE.
4. COVER THE VENT BASE WITH CEMENT.

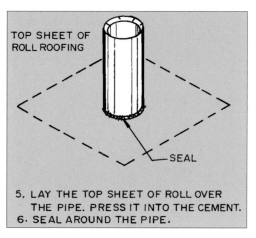

TOP SHEET OF ROLL ROOFING
SEAL

5. LAY THE TOP SHEET OF ROLL OVER THE PIPE. PRESS IT INTO THE CEMENT.
6. SEAL AROUND THE PIPE.

ADDITIONAL INFORMATION

Build-Up Roofing Design Guide for Building Owners, Asphalt Roofing Manufacturers Association, 6288 Montrose Rd., Rockville, MD 20852

Quality Control Guidelines for the Application of Built-Up Roofing and other publications, National Roofing Contractors Association, One O'Hare Centre, 6250 River Road, Rosemont, IL 60018

Roofing Specifications, Canadian Roofing Contractors Association, 116 Albert St., Suite 710, Ottawa, Ontario, Canada KIP 5G3

CHAPTER 11

Single-Ply Roofing

Single-ply roofing uses membrane products that are manufactured from synthetic materials and are very thin and flexible. A single thickness of membrane is used to finish the roof. Several types of single-ply roofing membrane are discussed in this chapter. Since there are variations in the products in each type manufactured, a choice of which type of membrane and the selection of a specific brand require the assistance of a qualified designer and a roofing contractor approved to install that particular membrane product.

While the homeowner ultimately makes the decision of which membrane to use and can oversee the contractor engaged to do the work, the do-it-yourselfer cannot apply this roofing alone and not every roofing contractor is qualified or approved to do the installation. Advice from the area sales office of the membrane manufacturer is also advised. Finally, be certain it is installed exactly as the manufacturer recommends and the fastening devices, cements, solvents, flashings, and other installation accessories provided by the manufacturer are used.

Check on the warranty from both the manufacturer and the roofing contractors.

SYNTHETIC RUBBER & PLASTIC SINGLE-PLY MEMBRANES

Some of the single-ply membranes in general use include ethylene propylene diene monomer (EPDM), chlorosulfoned polyethylene (CSPE), thermoplastic polyolefin (TPO), and polyvinyl chloride (PVC). These are all compounded synthetic materials. These thin flexible membranes are manufactured under controlled conditions, providing a uniform product. EPDM, CSPE, and TPO are referred to as **synthetic rubber**. PVC is a **thermoplastic** compound (11-1). There are other such products available, such as chlorinated polyethylene (CPE), polyisobutylene (PIB), and ethylene interpolymer (EIP). These will not be discussed.

11-1 The roof of this church (right) is being covered with a single-ply membrane (opposite) A variety of thermoplastic and thermoset roofing membranes are available, as discussed in the text.
Courtesy Johns Manville Roofing Systems

EPDM MEMBRANES

EPDM (ethylene propylene diene monomer) is a synthetic rubber membrane that is a thermoset plastic. It comes as a 45-mil- ($^{45}/_{1000}$ inch) and 60-mil- thick black sheet. Sheets are available from 10 to 100 feet wide. Most types are not fiber-reinforced, so they can be stretched a little, allowing them to lay over curved and irregularly shaped roofs. EPDM membrane resists many chemicals, but can be damaged by organic fats, gasoline, and fire. A specially formulated sheet that has fire-resistant properties is available.

The most commonly used type is black. It contains carbon black, enhancing its durability. White sheets do not have carbon black and have a shorter life. It can be coated with liquid Hypalon rubber available in several colors. CSPE membranes are made from Hypalon rubber; Hypan is the trade name of DuPont de Nemours.

EPDM membranes are available in rolls up to 50 feet wide and 100 feet long.

CSPE MEMBRANES

CSPE (chlorosulfoned polyethylene) is made as a thermoplastic; however, after it has been installed, it cures and becomes a thermoset. As mentioned just above, it is known by the trade name Hypalon. It is usually white. The sheet is reinforced with a polyester scrim. Scrim is a coarse, mesh-like material much like heavy cloth. Hypalon scrim is made with polyester fibers. This increases the tensile strength but reduces the ability to stretch. It is not as useful on unusually shaped roofs, where some elasticity is helpful.

Hypalon is generally white but other colors are available. It is resistant to many chemicals but can be damaged by gasoline and various oils. It is generally available in sheets 5 feet wide.

TPO MEMBRANES

TPO (thermoplastic polyolefin) is a membrane that has a high ethylene-propylene-rubber content. It is available in rolls of 75 inches, 8, 10, and 12 feet wide, and 100 feet long. It is available in black, white, and light gray, in 45- and 60-mil thicknesses (one mil is $^{1}/_{1000}$ of an inch). It resists UV and ozone exposure and extreme temperature variations. The seams are hot-air welded (**11-2**). It has fiber reinforcement, which increases its tear strength.

11-2 The seams of some types of single-ply membranes are sealed by a hot-air welding machine.
Courtesy Versico, Inc.

PVC MEMBRANES

PVC (polyvinyl chloride) is a thermoplastic sheet. It is a total plastic product and not considered in the synthetic rubber class as are EPDM, CSPE, and TPO. The PVC membranes do not have warranties that hold for as long as those offered for synthetic rubber membranes. Note that PVC can be damaged by some chemicals as well as oils and bitumens used in built-up roofs and asphalt shingles.

PVC membranes are available in gray, white, and several other colors (**11-3**). Since dirt tends to cling to these membranes, frequent cleaning is necessary if the roof is visible from the ground. They are available in rolls 75 inches wide and 100 feet long. Typical thicknesses are 48 mil to 100 mil.

INSTALLING SINGLE-PLY MEMBRANES

While the manufacturer will have specific installation recommendations, the following examples are typical of those available. Be certain to use the cements, adhesives, and other bonding techniques recommended. Flashing installations are also provided.

11-3 This roofing material is a single-ply membrane. It is being installed over an insulated roof deck.
Courtesy Johns Manville Roofing Systems

Synthetic rubber and thermoset, single-ply membranes are applied by three methods: **loosely laid ballasted**, **fully adhered**, and **mechanically fastened**. In **11-4**, a single-ply roofing membrane is being laid over an insulated roof deck; it can be installed in loosely laid ballasted, fully adhered, and mechanically fastened applications.

11-4 This single-ply membrane has been lapped over a parapet and the seam is being rolled.
Courtesy Versico, Inc.

LOOSELY LAID BALLASTED APPLICATION

Loosely laid membranes are laid over the roof sheathing and covered with a round, smooth rock or precast concrete pavers (**11-5**); this is the fastest and easiest installation method. When properly covered it could be given a UL Class A fire-resistance rating. The ballast adds to the roof load, so must be considered as the rafters are chosen. High winds can cause the ballast to move and the roof must have a low slope. A slope below 1:12 is recommended.

Thermoset single-ply membranes, such as EPDM, and CSPE are laid as shown in **11-6**.

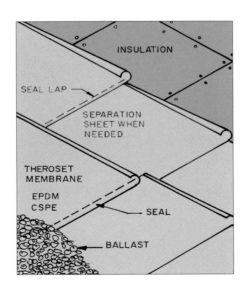

11-5 This single-ply membrane is loosely laid over the roof deck. After it has been attached to the perimeters and flashings, a layer of stone ballast is applied over the membrane. **Courtesy Versico, Inc.**

The insulation is either mechanically fastened to the sheathing or bonded with pads of adhesive. Some manufacturers recommend a separation layer be laid over the insulation if it has rough surfaces or sharp features that may damage the membrane.

After the membrane is laid over the insulation seal the edge and end laps with the recommended cement.

Hypalon (CSPE) membrane laps are welded together with a hot-air gun (refer to **11-3**, page 197). This softens the membrane, which is then rolled tight and fuses as it cools. A 3-inch lap is usually recommended. Carefully lay out the ballast so the roof has a uniform coating as recommended by the manufacturer.

FULLY ADHERED APPLICATION

The membrane can be adhered to the sheathing by coating the entire area with the manufacturer-recommended cement (**11-7**). The sheathing must be smooth. All dents, cracks, and splits should be filled. Coat the sheathing with the recommended primer.

If the roof has been covered with rigid insulation, attach it to the sheathing with mechanical fasteners.

11-6 Loose-laid thermoset membranes are rolled over the insulation. The edge and end laps are sealed and the surface is covered with a stone ballast.

11-7 The single-ply membrane is adhered to a manufacturer-supplied substrate adhesive. The seam is completed with a special primer, seam adhesives, and a special seam tape. **Courtesy Verico, Inc.**

11-8 A fully adhered membrane is bonded to the insulation by coating the insulation and back of the membrane with cement. The lap is cemented after the membrane has been laid.

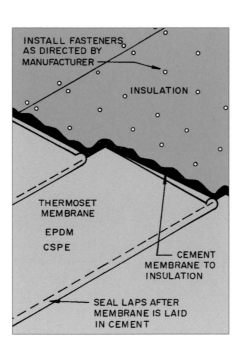

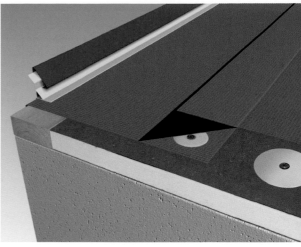

11-9 This mechanically attached membrane system is installed with manufacturer-supplied fasteners and batten bar.
Courtesy Versico, Inc.

Apply contact adhesive to the insulation or sheathing and the bottom of the membrane (unless you are using a self-adhering membrane). Begin by rolling the sheet out on the sheathing and aligning it with the eave or the previous sheet, allowing a 3-inch edge lap. Fold back half of the sheet and coat the sheathing and the bottom of the membrane. Do not coat the 3-inch edge lap. Allow the cement to set for the time specified. If the sheet is a self-adhering membrane remove the protective paper.

Now carefully lay the membrane over the coated sheathing, working to get a smooth surface. Work it with a large push-broom. Avoid getting wrinkles. Repeat for the other half. Then cement any laps with the approved cement (**11-8**).

MECHANICALLY ATTACHED APPLICATION

Mechanically attached membranes are loosely laid over the sheathing or insulation, spread out free of wrinkles, and secured to the sheathing with some type of mechanical fastener (**11-9**). Use the fastener supplied by the membrane manufacturer. A separation sheet is laid over the insulation or sheathing if it is rough and may damage the membrane.

Secure the membrane to the sheathing with one or another type of approved mechanical fastener (**11-10**). Two types frequently used are individual fasteners and bar fasteners. Space as recommended.

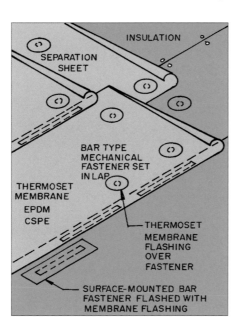

11-10 Membrane-roofing manufacturers have a variety of metal fasteners that penetrate the membrane and are secured to the sheathing. They are then covered with a layer of the same membrane used on the roof.

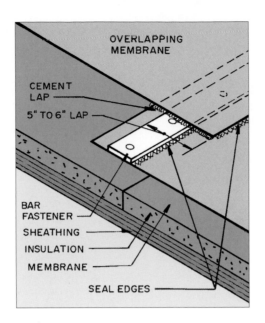

11-11 When metal fasteners are used along the edges of a membrane, a wide lap of the overlaying membrane seals them.

11-12 This single-ply-membrane roof on a commercial building has extensive flashing required to seal the many pipes and ducts that penetrate the roof deck.
Courtesy Johns Manville Roofing Systems

Then cover with pieces of membrane that are cemented to the roof membrane. Use the recommended cement; this seal is very important. Then seal the edge laps as recommended. In **11-11** a bar fastener has been placed in the edge lap and is sealed with a 6-inch overlay from the next membrane.

USING A THERMOPLASTIC MEMBRANE

Polyvinyl chloride (PVC) is the major thermoplastic membrane in use. PVC membranes can be laid using the fully adhered, mechanically attached, or the loose-laid ballasted method. The manufacturer's recommendations for laps, mechanical fasteners, and cements should be followed.

The loose-laid method will require that a separator sheet be laid over the roof sheathing. The membrane is laid over this as described for thermoset membranes. Edge laps should be at least 2 inches and end laps 4 inches. The laps are sealed by chemical fusion or heat-welded. Heat welding uses hot air to melt the membrane, which is then rolled together.

Mechanically attached membranes can be fastened with PVC-coated metal disks or bar-type fasteners as described for thermoset membranes. If PVC-coated metal disks are used, they are chemically welded or heat-welded to the roof membrane. The basic layout is the same as described for thermoset membranes.

Fully adhered membranes are installed the same as thermoset membranes. Use the adhesive recommended by the manufacturer. Typically water-based and solvent-based adhesives are used. After the adhesive has been applied to the sheathing and bottom of the membrane, let it set for the time required. Do not get the adhesive on the lap area, which will be sealed after the membrane has been laid and rolled smooth.

PVC flashing installation is about the same as described for thermoset flashing. Reinforced PVC flashing material is used. The roof membrane must be mechanically fastened to the sheathing in the area to be overlapped by the flashing. The flashing is chemically fused or heat-welded to the roof membrane. It is secured to the wall, chimney, or other butting surface with mechanical fasteners spaced 6 to 8 inches apart. The top edge is covered with counterflashing.

MODIFIED-BITUMEN MEMBRANES

Modified-bitumen membranes are sometimes installed as single-ply roofing; however, more likely they will be applied as multiple-ply roofing in a manner similar to built-up roofing, as discussed in Chapter 10. They are applied with hot bitumen. The installation details in Chapter 10 can be applied to single-ply or multiple-ply installation.

FLASHING

Membranes will terminate at places where flashing must be installed to seal the line of connection (**11-12**). The flashing may be cured or uncured elastomeric membrane material. An elastomeric membrane is one that has the elasticity to return to its original size after it has been stretched. Uncured elastomeric membranes cure after installation and are permanently set in the size and shape they assumed when installed. Thermoset membranes are cured and thermoplastic membranes are uncured.

Cured elastomeric membrane flashing is not used where it has to change direction and span a gap, such as where the roof butts the wall of the house. It should be the same material that was used on the roof.

Uncured elastomeric-membrane flashing can be used for any flashing job, including those requiring a change in direction and spanning a gap, as in the case where the roof butts a wall. After it has been installed, it will cure over time, permanently taking the shape in which it was laid.

Coat the bottom side of the flashing and the surface over which it will be laid with an approved cement. Let the cement set for the required time. Then lay the flashing on the roof and up over the wall or other butting area. Work to keep it smooth; a roller helps.

Suggestions for flashing at a masonry wall or chimney are offered in **11-13** and **11-14**. The top

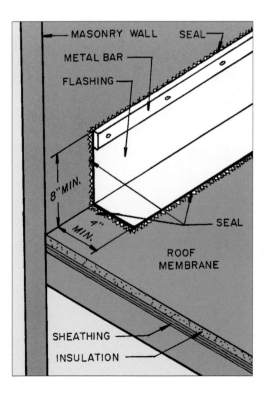

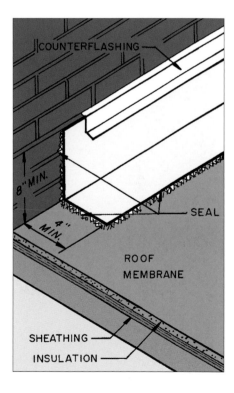

11-13 (Left)
The membrane flashing is bonded to the masonry wall and roof membrane. The top is secured with manufacturer-supplied metal bars. The edges are sealed.

11-14 (Right)
This membrane is bonded to the brick wall and roof membrane. The top is sealed with counterflashing.

11-15 After this membrane has been bonded to the wall and roof membrane, the wood siding is extended over it, sealing the top edge.

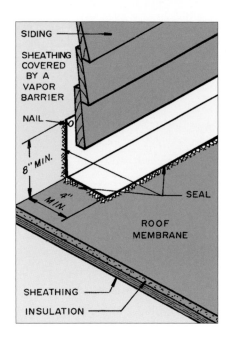

edge may be secured with counterflashing or a metal strip. One way to flash this intersection if a wood wall is involved is shown in **11-15**. The siding overlaps the flashing, sealing the top edge.

An outside corner is flashed by first flashing the wall parallel with the eave. Then flash the side wall and fold the flashing around the corner and cement to the front flashing, roof membrane and wall (**11-16**).

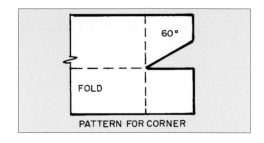

11-16 (Left) Steps to flash an outside corner where a roof meets a wall.

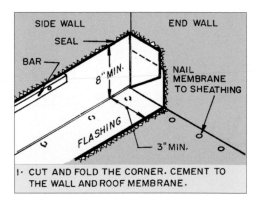

11-17 (Right top, middle, and bottom) Steps to flash an inside corner where a roof meets a wall.

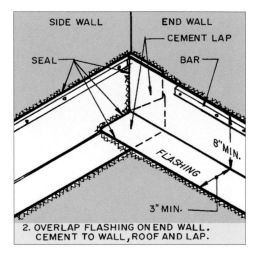

An inside corner is flashed by first installing the flashing on the wall parallel with the eave and overlapping it with the side flashing (11-17).

Pipe penetrations are sealed by installing a plastic pipe flashing unit over the pipe. The manufacturer will have products available. A typical example is shown in **11-18**.

There are various ways to flash the eave. Typical details are shown in **11-19** and **11-20**. The roof membrane is laid over the edge of the roof and secured to the fascia. Metal flashing is laid over this and its edge is sealed with a layer of flashing membrane.

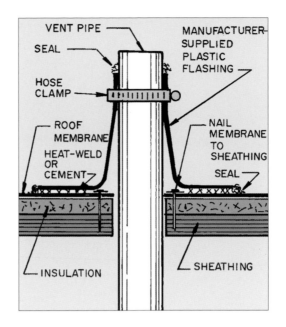

11-18
A generalized example of plastic flashing to secure a vent pipe. Use the unit supplied by the manufacturer and install as directed.

11-19 (Left) One way to flash the eave when the roof is finished with a single-ply membrane.

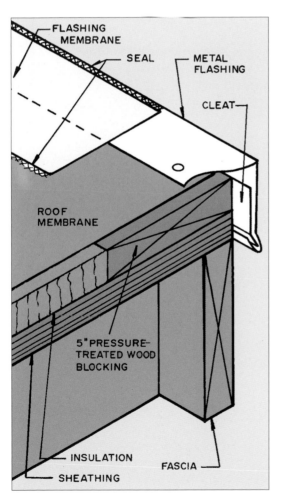

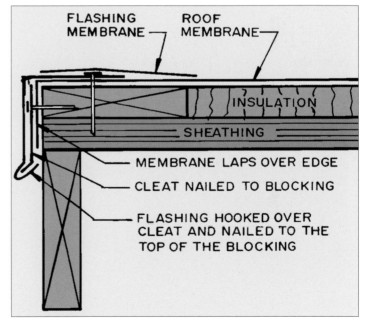

11-20 (Above) A section through the eave showing the relationship between the cleat and the metal edge flashing. The cleat holds the edge flashing in place during high winds and allows it to be free of exposed nails.

11-21 A general example of the finished installation of a single-ply-membrane roofing material

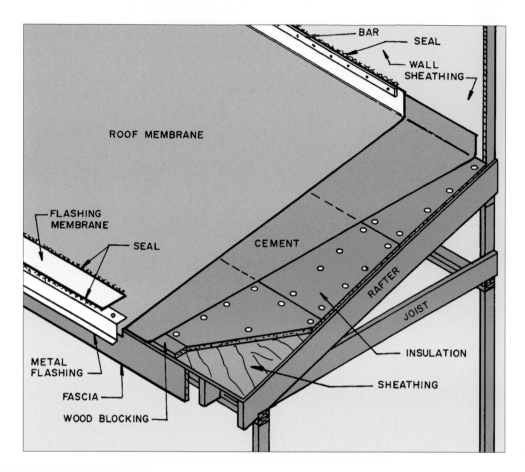

An overall example of the total single-ply membrane roofing installation and flashing can be seen in **11-21**.

Index

Algae, 55, 92–93
Asphalt shingle installation, 75–87
  chalk lines, 81–82
  first course, 81–83
  flashing, 88–92
  in high wind areas, 78
  interlocking shingles, 86–87
  on low-slope roofs, 76–78
  nailing, 75, 78, 83
  patterns, 78–81
  reroofing steps, 45–51
  ridge/hip capping, 84–85
  on steep-slope roofs, 83
  valley flashing, 88–90
Asphalt shingles, 72–93
  algae discoloration control,
    92–93
  features, 11, 73
  inorganic type, 11
  installing. See Asphalt shingle
    installation
  organic type, 11
  over wood shingles, 49
  rafter inspection, 46
  repairing, 33–36
  replacing individual, 34–35
  self-sealing, 73–74
  slope range for, 9
  storing, 23
  styles, 73–74
  tearing off old, 51–52
  weight, 117
  wood shingles over, 50–51

Battens, 119–121
Bitumens. See Built-up roofing
  (BUR); Modified-bitumen
  membranes
Built-up roofing (BUR), 177–185
  base sheets, 179
  cold-applied bitumens, 178
  considerations, 177
  features, 177
  flashing, 183–185

hot bitumens, 177–178
insulation board products, 179
membrane, 179–181
modified-bitumen membranes,
  9, 13, 177, 182–183
nailing, 180–181
roofing felts, 178–179
sheathing, 177
slope range for, 9
surface aggregates, 179
typical construction, 178,
  179–181
Butterfly roofs, 10

Cathedral ceiling ventilation,
  30–31
Cedar shakes/shingles. See Wood
  shakes/shingles
Cement-fiber shakes/shingles, 7, 12
Chalk lines, 81–82, 123
Chlorosulfoned polyethylene
  (CSPE) membranes,
  195–196
Clay & concrete tile, 114–147
  additional resources, 147
  barrel (mission) tile, 116, 119,
    132–135
  battens for, 119–121
  cutting, 136–137
  features, 12, 115–116
  flat tile, 117, 119, 123–126
  installing. See Clay & concrete
    tile installation
  look-alikes, 7
  low-profile tile, 117, 126–130,
    139–140
  preparation, 119–122
  profile options, 116–117
  safety, 122
  sheathing, 118
  slope range for, 9
  snow guards, 137
  spaced sheathing boards, 42–43
  S-tile, 116, 117, 119, 130–132

storing, 23
tearing off old roofs, 53
underlayment, 118
weight, 117
Clay & concrete tile installation,
  123–147
  assistance with, 122
  barrel (mission) tile, 132–135
  chalk lines, 123
  flashing, 138–147
  flat tile, 123–126
  low-profile tile, 126–130,
    139–140
  nails/nailing, 123
  S-tile, 130–132
Coal-tar built-up roofs. See Built-
  up roofing (BUR)
Combing slates, 106–108
Comb of roof, 106
Comb ridge, 107–108
Composite-board insulation, 179
Copper nails, 100
Crickets, 67

Drip edge, 75

Estimating materials, 23–26
  calculating squares, 23–24
  figuring roof area, 24–25
  flashing, 26
  ridge/hip caps, 26
  shingle requirements, 24–25
  waste allowance, 25
Ethylene propylene diene
  monomer (EPDM)
  membranes, 195–196
Expanded perlite insulation board,
  179
Eye protection, 15

Fiberboard panel insulation, 179
Fiberboard sheathing, 41–43
Fire resistance, 6

Flashing
    asphalt shingles, 88–92
    built-up roofing, 183–185
    clay & concrete tile, 138–147
    estimating needs, 26
    metal roofing, 167–172
    repairing, 39
    roll roofing, 191–193
    single-ply roofing, 201–204
    slate roofing, 108–112
    wood shakes/shingles, 63–69
Fungi, 55

Gable roofs, 9–10, 27–29
Galvanized steel nails, 100
Gambrel roofs, 10

Hip cap estimates, 26
Hip roofs, 9–10

Insulation board products, 179
Isosceles triangles, 24–25

Knee pads, 15

Ladder jacks, 21–22
Ladder safety, 17–18
Lath (open) sheathing, 42–43,
    97–98, 106
Leak repair, 33
Lifting safety, 15–16, 22

Mansard roofs, 10
Materials. See also specific materials
    estimating, 23–26
    features overview, 10–13
    roofing options, 7–8
    storing, 23, 41
    Metal roofing, 148–175
    coatings, 150–151
    features, 13, 149–150
    galvanic corrosion, 150
    installing. See Metal roofing
        installation
    metal options, 154–156
    panels, 152–154, 159–161,
        162–165
    shingles, 172–175
    simulating materials, 7, 172–174

slope range for, 9
storing, 23
tearing off old roofs, 53
thermal considerations,
    151–152, 157
types/profiles, 152–156,
    172–175
ventilating, 156
Metal roofing installation,
    156–175
    cutting panels, 158
    fastening panels, 159–161
    flashing, 167–172
    installing panels, 162–165
    panel end laps, 165
    ridge/hip capping, 166–167
    sheathing, 157–158
    shingles, 172–175
Mineral-surfaced roll roofing. See
    Roll roofing
Modified-bitumen membranes, 9,
    13, 177, 182–183, 201
Moisture control, 26–27
Monitor shed roofs, 10
Moss, 55

Nails/nailing, 41–42
    asphalt shingles, 75, 78, 83
    built-up roofing, 180–181
    clay & concrete tile, 123
    roll roofing, 186–190
    slate roofing, 100–101
    wood shakes/shingles, 60–61

OSB (oriented strandboard)
    sheathing, 41
OSHA (Occupational Safety and
    Health Administration)
    requirements, 16–18, 19, 20

Perm ratings, 27
Pitch, 8
Plywood sheathing, 41
Polyisocyanurate (ISO) insulation
    board, 179
Polyvinyl chloride (PVC) mem-
    branes, 195, 197, 200
Power lifts, 22
Pump jacks, 20

Rafter inspection, 46
Repairs, 32–39
    asphalt shingles, 33–36
    flashing, 39
    roof leaks, 33
    slate roofs, 38–39, 112
    wood shingles/shakes, 36–37
Reroofing
    asphalt over asphalt, 45–49
    asphalt over wood, 49
    building codes, 45
    decision considerations, 45
    tearing off old roofs, 51–53
    wood over asphalt, 50–51
Ridge cap estimates, 26
Ridge vents, 27, 29–31, 66,
    112–113
Right triangles, 24–25
Rise, 8
Roll roofing, 185–193
    common types, 185
    features, 13, 177, 185
    flashing, 191–193
    installing, 186–193
    mineral-surfaced selvedge-edge,
        189–190
    nails/nailing, 186–190
    ridge/hip capping, 191
    slope range for, 9
Roof brackets, 21, 104
Roof slopes, 8–10
    appropriate materials, 9
    defined, 8
    low-sloped, 9
    OSHA safety requirements,
        16–17
    steep-sloped, 9
Roof types, 9–10

Saddle hip, 108
Saddle ridge, 107
Safety, 15–22
    clay & concrete tile, 122
    eye protection, 15
    knee pads, 15
    ladder jacks, 21–22
    ladders, 17–18
    lifting, 15–16, 22
    OSHA regulations, 16–20
    personal, equipment, 15–16

power lifts, 22
pump jacks, 20
roof brackets, 21
scaffolding, 19, 102–104
shingle pads, 14–15
Scaffolding, 19, 102–104. *See also* Ladder jacks
Selecting roofing, 4–13. *See also specific materials*
  aesthetic considerations, 5
  fire resistance and, 6
  roof slope and, 8–10
  warranty considerations, 7
  weight considerations, 8
  wind resistance ratings, 7
Shakes. *See* Wood shakes/shingles
Sheathing, 41–43
  built-up roofing, 177
  clay & concrete tile, 118
  lath boards, 97–98
  materials, 41–43
  metal roofing installation, 157–158
  open (lath), 42, 43, 97–98, 106
  slate roofing, 97–98
  wood shakes/shingles, 58
Shed roofs, 9–10
Shingle estimates, 24–25
Shingle pads, 14–15
Shingles. *See specific materials*
Single-ply roofing, 194–203
  features, 195
  flashing, 201–204
  installing, 197–204
  modified-bitumen membranes, 201
  slope range for, 9
  synthetic rubber/plastic, 195–197
Slate roofing, 94–113
  additional resources, 113
  cutting, 101–102
  exposure levels, 98
  features, 11, 95

graduated, 96
hole punching, 98–100
installing. *See* Slate roofing installation
look-alikes, 7
material/size options, 95–96
nails/nailing, 100–101
preparation, 96–102
repairing/replacing slates, 38–39, 112
sheathing, 97–98
slope range for, 9
spaced sheathing boards, 42–43
standard, 96
tearing off old roofs, 53
textural, 96
Slate roofing installation, 102–112
  combing slates, 106–108
  flashing, 108–112
  on open sheathing, 106
  ridge/hip capping, 98, 106–108
  ridge vents, 112–113
  scaffolding for, 102–104
  on solid sheathing, 104–106
Slate-type roofing, 9
Slopes. *See* Roof slopes
Snow guards, 137
Soffit vents, 27, 29–31, 66
Span, 8
Sprayed-polyurethane-foam roofing, 9
Squares, estimating, 23–25
Staples, 60–61, 78
Storing materials, 23, 41
Strip-saddle ridge, 107–108
Synthetic rubber/plastic. *See* Single-ply roofing

Tearing off roofs, 51–53
Thermoplastic compounds, 195–197, 200
Thermoplastic polyolefin (TPO) membranes, 195–196
Triangles, 24–25

Underlayment, 44–45
  for clay & concrete tile, 118
  estimating needs, 24–25
  inorganic felts, 44
  installing, 75–78
  on low-slope roofs, 76–78
  organic felts, 44–45, 75
  synthetic resins, 45

Ventilation, 26–31
  cathedral ceilings, 30–31
  gable roofs, 27–29
  metal roofing and, 156
  mixing types of, 31
  overview, 26–27
  perm ratings, 27
  requirements, 26–27

Waferboard sheathing, 41
Warranties, 7, 55
Waste allowance, 25
Weight considerations
  asphalt shingles, 117
  clay & concrete tile, 117
  selecting materials, 8
  stacking shingles on roof, 15
Wind resistance ratings, 7
Wood shakes/shingles, 55–71
  cedar shakes, 57–58
  cedar shingles, 56, 58
  composite/recycled, 8, 71
  features, 11, 55
  flashing, 63–69
  installing shakes, 69–71
  installing shingles, 58–63
  over asphalt shingles, 50–51
  recommended exposure, 58, 61
  replacing individual, 36–37
  sheathing, 58
  slope range for, 9
  spaced sheathing boards, 42–43
  storing, 23
  tearing off old roofs, 53
  warranties, 55

Metric Equivalents

inches	mm	cm	inches	mm	cm	inches	mm	cm
⅛	3	0.3	13	330	33.0	38	965	96.5
¼	6	0.6	14	356	35.6	39	991	99.1
⅜	10	1.0	15	381	38.1	40	1016	101.6
½	13	1.3	16	406	40.6	41	1041	104.1
⅝	16	1.6	17	432	43.2	42	1067	106.7
¾	19	1.9	18	457	45.7	43	1092	109.2
⅞	22	2.2	19	483	48.3	44	1118	111.8
1	25	2.5	20	508	50.8	45	1143	114.3
1¼	32	3.2	21	533	53.3	46	1168	116.8
1½	38	3.8	22	559	55.9	47	1194	119.4
1¾	44	4.4	23	584	58.4	48	1219	121.9
2	51	5.1	24	610	61.0	49	1245	124.5
2½	64	6.4	25	635	63.5	50	1270	127.0
3	76	7.6	26	660	66.0			
3½	89	8.9	27	686	68.6	inches	feet	m
4	102	10.2	28	711	71.1			
4½	114	11.4	29	737	73.7	12	1	0.31
5	127	12.7	30	762	76.2	24	2	0.61
6	152	15.2	31	787	78.7	36	3	0.91
7	178	17.8	32	813	81.3	48	4	1.22
8	203	20.3	33	838	83.8	60	5	1.52
9	229	22.9	34	864	86.4	72	6	1.83
10	254	25.4	35	889	88.9	84	7	2.13
11	279	27.9	36	914	91.4	96	8	2.44
12	305	30.5	37	940	94.0	108	9	2.74

Conversion Factors

						mm	=	millimeter
1 mm	=	0.039 inch	1 inch	=	25.4 mm	cm	=	centimeter
1 m	=	3.28 feet	1 foot	=	304.8 mm	m	=	meter
1 m^2	=	10.8 square feet	1 square foot	=	0.09 m^2	m^2	=	square meter